ASAP
Biology

By the Staff of The Princeton Review

princetonreview.com

Penguin
Random
House

The Princeton Review
110 East 42nd Street, 7th Floor
New York, NY 10017
Email: editorialsupport@review.com

Copyright © 2017 by TPR Education IP
Holdings, LLC. All rights reserved.

Published in the United States by Penguin
Random House LLC, New York, and in Canada
by Random House of Canada, a division of
Penguin Random House Ltd., Toronto.

Terms of Service: The Princeton Review Online
Companion Tools ("Student Tools") for retail
books are available for only the two most recent
editions of that book. Student Tools may be
activated only twice per eligible book purchased
for two consecutive 12-month periods, for a
total of 24 months of access. Activation of
Student Tools more than twice per book is in
direct violation of these Terms of Service and
may result in discontinuation of access to
Student Tools Services.

ISBN: 978-1-5247-5764-9
eBook ISBN: 978-1-5247-5769-4
ISSN: 2573-8070

AP and Advanced Placement are registered
trademarks of the College Board, which is not
affiliated with The Princeton Review.

The Princeton Review is not affiliated with
Princeton University.

Editor: Sarah Litt
Production Editor: Jim Melloan
Production Artists: Deborah A. Silvestrini and
            Craig Patches

Printed in the United States of America.

10 9 8 7 6 5 4 3 2 1

Editorial

Rob Franek, Editor-in-Chief
Casey Cornelius, VP Content Development
Mary Beth Garrick, Director of Production
Selena Coppock, Managing Editor
Meave Shelton, Senior Editor
Colleen Day, Editor
Sarah Litt, Editor
Aaron Riccio, Editor
Orion McBean, Associate Editor

Penguin Random House Publishing Team

Tom Russell, VP, Publisher
Alison Stoltzfus, Publishing Director
Jake Eldred, Associate Managing Editor
Ellen Reed, Production Manager
Suzanne Lee, Designer

Acknowledgments

The Princeton Review would like to thank Jes Adams and Nathan Chamberlain for all their work on this book.

Jes Adams would like to thank Katherine Wright, Michelle Axford, and Neil Adams.

Nathan Chamberlain would like to thank the fantastic staff at The Princeton Review for all their guidance, my family for raising me to drive onward through the toughest challenges, my AP Biologists for showing me what they really needed, and most importantly, my wonderfully brilliant wife, Katie, for the hours of suggestions and support, as well as her tolerance of all of my idiosyncrasies.

The editor would also like to give special thanks to Debbie Silvestrini and Craig Patches for their patience and hard work. This book would not be as lovely as it is if it were not for them.

Contents

Get More (Free) Content

1 Go to **PrincetonReview.com/cracking.**

2 Enter the following ISBN for your book: 9781524757649.

3 Answer a few simple questions to set up an exclusive Princeton Review account. (If you already have one, you can just log in.)

4 Click the "Student Tools" button, also found under "My Account" from the top toolbar. You're all set to access your bonus content!

Need to report a potential **content** issue?

Contact
EditorialSupport@review.com.
Include:
- full title of the book
- ISBN number
- page number

Need to report a **technical** issue?

Contact
TPRStudentTech@review.com
and provide:
- your full name
- email address used to register the book
- full book title and ISBN
- computer OS (Mac/PC) and browser (Firefox, Safari, etc.)

Once you've registered, you can...

- Get valuable advice about the college application process, including tips for writing a great essay and where to apply for financial aid

- If you're still choosing between colleges, use our searchable rankings of *The Best 382 Colleges* to find out more information about your dream school

- Access a variety of printable resources, including bonus "could know" material and additional charts

- Check to see if there have been any corrections or updates to this edition

- Get our take on any recent or pending updates to the AP Biology Exam

Introduction

What is This Book and When Should I Use It?

Welcome to *ASAP Biology,* your quick-review study guide for the AP Exam written by the Staff of The Princeton Review. This is a brand-new series custom built for crammers, visual learners, and any student doing high-level AP concept review. As you read through this book, you will notice that there aren't any practice tests, end-of-chapter drills, or multiple-choice questions. There's also very little test-taking strategy presented in here. Both of those things (practice and strategy) can be found in The Princeton Review's other top-notch AP series—*Cracking.* So if you need a deep dive into AP Biology, check out *Cracking the AP Biology Exam* at your local bookstore.

ASAP Biology is our fast track to understanding the material—like a fantastic set of class notes. We present the most important information that you MUST know (or should know or could know—more on that later) in visually friendly formats such as charts, graphs, and maps, and we even threw a few jokes in there to keep things interesting.

Use this book any time you want—it's never too late to do some studying (nor is it ever too early). It's small, so you can take it with you anywhere and crack it open while you're waiting for soccer practice to start or for your friend to meet you for a study date or waiting for the library to open. *ASAP Biology* is the perfect study guide for students who need high-level review in addition to their regular review and also for students who perhaps need to cram pre-Exam. Whatever you need it for, you'll find no judgment here!

 Because you camp out in front of the library like they are selling concert tickets in there, right? Only kidding.

Who is This Book For?

This book is for YOU! No matter what kind of student you are, this book is the right one for you. How do you know what kind of student you are? Follow this handy chart to find out!

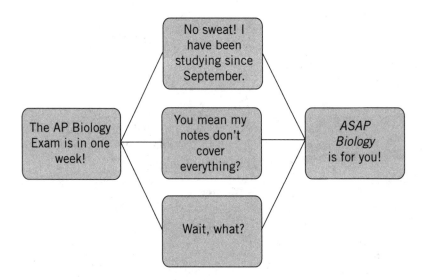

As you can see, this book is meant for every kind of student. Our quick lessons let you focus on the topics you must know, you should know, and you could know—that way, even if the test is tomorrow (!), you can get a little extra study time in, and only learn the material you need.

How Do I Use This Book?

This book is your study tool, so feel free to customize it in whatever way makes the most sense to you, given your available time to prepare. Here are some suggestions:

Target Practice

If you know what topics give you the most trouble, hone in on those chapters or sections.

ASK Away

Answer all of the ASK questions *first*. This will help you to identify any additional tough spots that may need special attention.

Three-Pass System

Start at the very beginning!* Read the book several times from cover to cover, focusing selectively on the MUST content for your first pass, the SHOULD content for your second pass, and finally, the COULD content.

 *It's a very good place to start.

Why Are There Icons?

Your standard AP course is designed to be equivalent to a college-level class, and as such, the amount of material that's covered may seem overwhelming. It's certainly admirable to want to learn everything—these are, after all, fascinating subjects. But every student's course load, to say nothing of his or her life, is different, and there isn't always time to memorize every last fact.

To that end, *ASAP Biology* doesn't just distill the key information into bite-sized chunks and memorable tables and figures. This book also breaks down the material into three major types of content:

❶ This symbol calls out a section that has MUST KNOW information. This is the core content that is either the most likely to appear in some format on the test or is foundational knowledge that's needed to make sense of other highly tested topics.

⬤ This symbol refers to SHOULD KNOW material. This is either content that has been tested in some form before (but not as frequently) or which will help you to deepen your understanding of the surrounding topics. If you're pressed for time, you might just want to skim it, and read only those sections that you feel particularly unfamiliar with.

◑ This symbol indicates COULD KNOW material, but don't just write it off! This material is still within the AP's expansive curriculum, so if you're aiming for a perfect 5, you'll still want to know all of this. That said, this is the information that is least likely to be directly tested, so if the test is just around the corner, you should probably save this material for last.

As you work through the book, you'll also notice a few other types of icons.

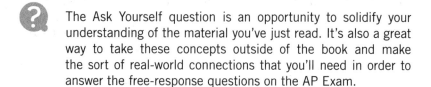

The Ask Yourself question is an opportunity to solidify your understanding of the material you've just read. It's also a great way to take these concepts outside of the book and make the sort of real-world connections that you'll need in order to answer the free-response questions on the AP Exam.

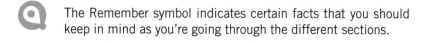

The Remember symbol indicates certain facts that you should keep in mind as you're going through the different sections.

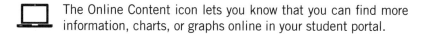

The Online Content icon lets you know that you can find more information, charts, or graphs online in your student portal.

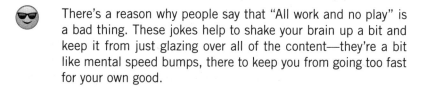

There's a reason why people say that "All work and no play" is a bad thing. These jokes help to shake your brain up a bit and keep it from just glazing over all of the content—they're a bit like mental speed bumps, there to keep you from going too fast for your own good.

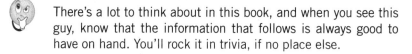

There's a lot to think about in this book, and when you see this guy, know that the information that follows is always good to have on hand. You'll rock it in trivia, if no place else.

Where Can I Find Other Resources?

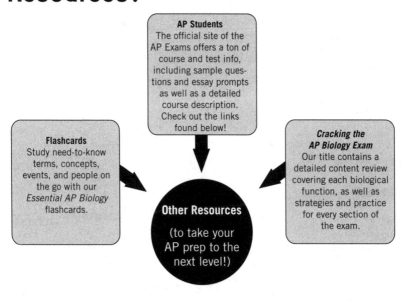

AP Students
The official site of the AP Exams offers a ton of course and test info, including sample questions and essay prompts as well as a detailed course description. Check out the links found below!

Flashcards
Study need-to-know terms, concepts, events, and people on the go with our *Essential AP Biology* flashcards.

Cracking the AP Biology Exam
Our title contains a detailed content review covering each biological function, as well as strategies and practice for every section of the exam.

Other Resources
(to take your AP prep to the next level!)

Useful Links

- AP Biology Homepage: https://apstudent.collegeboard.org/apcourse/ap-biology
- Your Student Tools: www.PrincetonReview.com/cracking
  See the "Get More (Free) Content" page for step-by-step instructions for registering your book and accessing more materials to boost your test prep.

CHAPTER 1

Chemistry of Life

All living things are made of cells and cells are made of molecules. Water makes life possible, and organisms are made of mostly water. Cells also contain four types of biological macromolecules that play important roles in supporting life.

Organization of Life ∼

Organization in living things exists in a hierarchical structure.

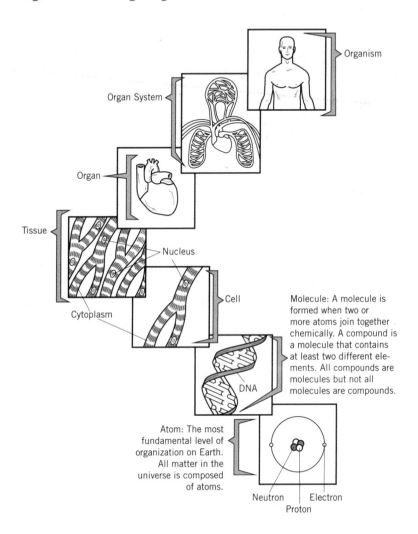

Organism

Organ System

Organ

Tissue

Nucleus

Cell

Cytoplasm

DNA

Molecule: A molecule is formed when two or more atoms join together chemically. A compound is a molecule that contains at least two different elements. All compounds are molecules but not all molecules are compounds.

Atom: The most fundamental level of organization on Earth. All matter in the universe is composed of atoms.

Neutron | Electron
Proton

~ Atoms are made of subatomic particles:

Subatomic Particle	Location	Charge	Size	Mass
Proton	Nucleus	+1	Large	Large
Neutron	Nucleus	0	Large	Large
Electron	Orbiting around the nucleus	−1	Tiny	Tiny

The building blocks of life (atoms, small molecules, and biological macromolecules like proteins) cannot be seen with the naked eye or a light microscope. You would need an electron microscope to see them. Organisms vary considerably in size. Here are some members of the biological kingdom, and how they relate to one another in size.

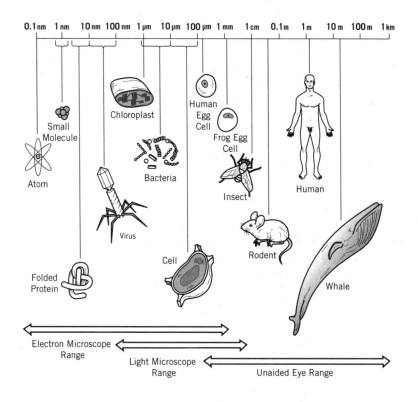

 "Large" is a relative term here. The radius of a proton is about a femtometer, or about 1×10^{-15} meters.

Essential Elements of Life

An element is a substance that is made entirely from one type of atom. These elements are essential for life:

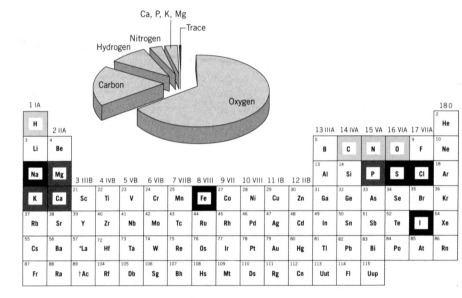

By mass, living things are:

- 96% O, C, H, and N (light gray above)
- 3.5% Ca, P, K, and Mg (dark gray above)
- 0.5% other trace elements (Na, Cl, S, Fe, I, etc.) (black above)

Ask Yourself...

1. Which subatomic particles are different in isotopes of an element? What about in ions?

2. Which of our physiological systems make sure each of our cells get the oxygen they require to survive?

Oxygen

Final electron acceptor in cell respiration and made in photosynthesis

Phosphorus

Used in nucleic acids and some lipids

Carbon

- Backbone of organic molecules

- Found in carbohydrates, proteins, lipids, and nucleic acids

Hydrogen

A small element found in most organic molecules

Nitrogen

Used in proteins and nucleic acids

Water, H$_2$O ❗

Living systems depend on 4 chemical properties of water:

1. Water is a versatile solvent
2. Water molecules are cohesive and adhesive
3. Water expands when it freezes
4. Water can moderate temperature changes

Water is a Versatile Solvent ❗

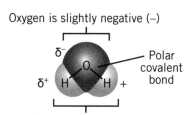

Oxygen is slightly negative (−)

δ^-

O

δ^+ H   H +

Polar covalent bond

Hydrogens are slightly positive (+)

Water is polar because oxygen is more electronegative than hydrogen

---

Because it is polar, water is the universal solvent:

- Can dissolve many other polar molecules
- Supports reactions in our cells and our environment
- Living systems depend on water
- Hydrophilic substances have an affinity for water

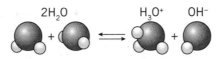

$2H_2O$ H_3O^+ OH^-

- Water can act as both proton donor (acid) and proton acceptor (base), so it is **amphoteric**
- pH of pure water is 7.0

 Ask Yourself...

What kinds of compounds would a hydrophobic molecule be attracted to?

Water Molecules are Cohesive and Adhesive ❗

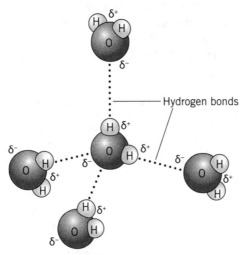

Hydrogen bonds

Water molecules can hydrogen bond:

- A type of intermolecular interaction (weak chemical bond between two molecules)
- Strong interactions when present in large numbers
- Water can form a maximum of four hydrogen bonds with other molecules

 Water accounts for about 70% of an animal cell by mass.

Chemistry of Life

Water molecules are **cohesive**: they stick together because of hydrogen bonding

Implications of water being cohesive:

- Moving water molecules can pull on their neighbor molecules; this contributes to transpiration in plants
- Water has a high surface tension
- Water forms droplets on many surfaces

Water molecules are **adhesive:** they are attracted to other molecules

Implications of water being adhesive:

- Water can move up a narrow tube because of adhesion; this is called capillary action and contributes to transpiration in plants
- Water has a concave up meniscus

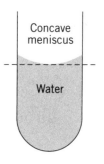

 Ask Yourself...

What kind of molecules have a convex meniscus?

 Insects called water striders can "walk" on the surfaces of ponds because the surface tension of water is high.

ASAP Biology

Water Expands When it Freezes

- The freezing point of water is 0°C / 32°F
- Water molecules spread out when they freeze because of hydrogen bonds

Implications:

- Ice floats on the top of lakes or rivers, and animals can live underneath, in liquid water
- Pipes and bottles containing water can explode when they freeze

❓ Ask Yourself...

Is ice more or less dense than liquid water? What about steam?

 About 70% of the Earth is covered in water but only 3% of this is fresh water (in ice, rivers and lakes). The rest is salt water in oceans.

Water Can Moderate Temperature Changes

- Water vaporizes at a very high heat
- It takes a large amount of heat for liquid water to become a gas (because all hydrogen bonds must be broken)
- The boiling point of water is 100°C / 212°F

Application: sweating cools us down because heat is used to evaporate sweat (which is 90% water), leaving your body cooler.

- Water has a high heat capacity and a high specific heat
- Water can absorb or lose a large amount of heat and not vary temperature much
- Water resists temperature changes

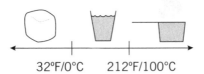

32°F/0°C   212°F/100°C

Applications:
- Water absorbs heat from warm air and releases heat to cool air
- Pool, lake, and ocean water is still warm in the evening when the air is chilly
- Organisms can maintain a constant body temperature

The average human is about 50–70% water, depending on gender, age, and fitness level.

Biological Macromolecules

All biological macromolecules are polymers, or strings of repeated units (monomers). They are grouped into four classes of molecules and play important roles in cells and organisms.

Building Blocks and Bonds				
Organic Molecule	Monomer	Two monomers	Many monomers	Bond
Carbohydrates	Monosaccharide	Disaccharide	Polysaccharide	Glycosidic bond
Proteins	Amino acid	Dipeptide	Polypeptide (unfolded) Protein (folded)	Peptide bond
Lipids	Uses 3 building blocks instead: 1. Glycerol 2. Fatty acids 3. Four rings of carbon	n/a	Triglyceride Phospholipid Cholesterol Steroids	Ester linkage
Nucleic acids	Nucleotides	Dinucleotide	RNA DNA	Sugar-phosphate phosphodiester bonds

Dehydration Synthesis

- Forms covalent bonds
- Water is released
- Also called condensation
- Examples:
    - Connect amino acids → form peptide bond
    - Connect monosaccharides → form glycosidic bond
    - Connect fatty acids to glycerol → form ester linkage

Example 1: Peptides

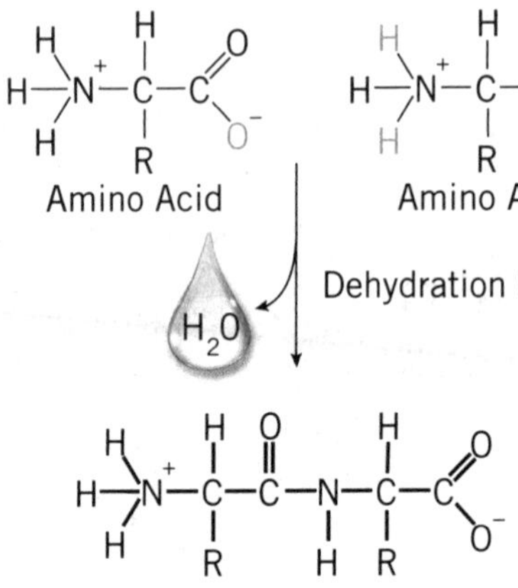

Amino Acid   Amino Acid

Dehydration Synthesis

H_2O

Dipeptide

Example 2: Carbohydrates

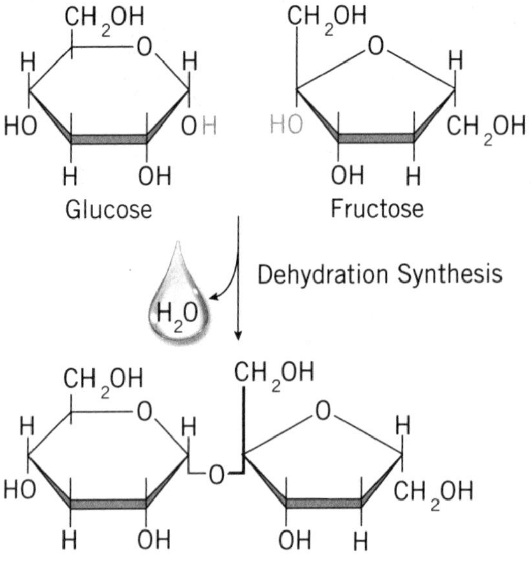

Glucose   Fructose

Dehydration Synthesis

H_2O

Sucrose

Carbohydrates ❗

Carbohydrates are energy-storing organic compounds that contain the elements carbon, oxygen, and hydrogen; that's where the name is derived from. Carbohydrates also form many structural units (especially in plants) and help with cell-cell attachment. Carbohydrates are also called sugars, or saccharides.

Monosaccharides ❗

- Single sugar monomers are called monosaccharides
- They can be in a ring or straight-chain form
- There is always a ratio of 1:2:1 of carbon, oxygen, and hydrogen atoms, so their formula is always $C_nH_{2n}O_n$
- Glucose, fructose, and galactose are monosaccharides

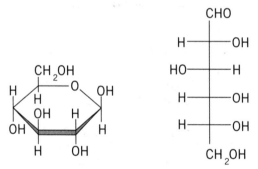

Glucose (ring form)   Glucose (straight-chain form)

Disaccharides ❗

- Monosaccharides can be linked together by dehydration synthesis to form disaccharides
- This forms a glycosidic linkage between the two sugar monomers
- Maltose, lactose, and sucrose are disaccharides

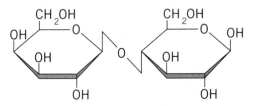

Lactose

Polysaccharides ❗

- When more than two sugars connect, they are called polysaccharides
- Starch, glycogen, and cellulose are polysaccharides
- We can easily digest glycogen, but we don't have the enzymes to breakdown cellulose
- Each hexagon in this diagram represents a monosaccharide

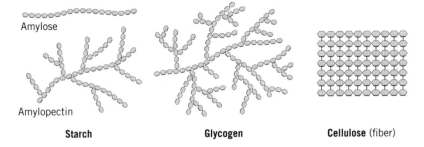

Amylose

Amylopectin

Starch   **Glycogen**   **Cellulose** (fiber)

Maltose is the building block of starch, lactose is found in milk, and sucrose is table sugar.

Lactose has a different type of glycosidic bond than maltose and sucrose. This is why about 25% of Americans are lactose intolerant: they don't make the enzyme to break down this type of bond.

Proteins ❗

Proteins have many jobs in the cell: they help transport things around, provide structural support, provide defence, speed up reactions, provide energy, and help with energy storage.

Amino Acids ❗

Proteins are made of amino acids. Every amino acid has a central carbon atom with:

1. Amino group—basic
2. Carboxyl group—acidic
3. R group or side chain
4. Hydrogen

Amino Acid

$$H_3N^+ - C - C \diagdown O \diagup OH$$

Amino Group   "R" Group   Carboxyl Group

Amino acids always contain:

1. Carbon
2. Nitrogen
3. Oxygen
4. Hydrogen

Animals store glucose in glycogen and plants store glucose in starch. This is why you can eat and digest potatoes and rice, but not your T-shirt.

Two amino acids also contain sulfur.

There are 20 amino acids and each one has a different R-group. These side chains can be hydrophobic or hydrophilic. Depending on the chemical properties of each side chain, the amino acid can have very diverse characteristics.

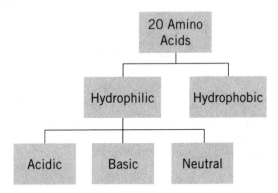

Ask Yourself...

What elements or groups make an R group polar? Non-polar?

Protein Structure ❗

- To form a protein, amino acids are connected together by peptide bonds
  - Dipeptide: 2 amino acids bound together
  - Polypeptide: many amino acids
- Ends of a peptide chain are called N-terminus and C-terminus
  - New amino acids are always added onto the C-terminus in the cell
- Disulfide linkages are covalent bonds between cysteine amino acids

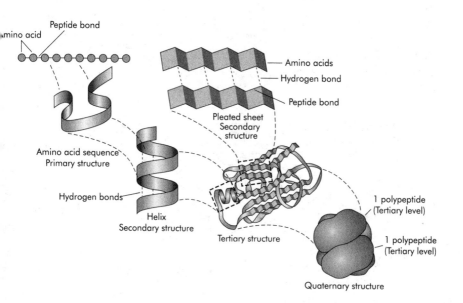

Here is a summary of protein structure:

Level	What it is	Held together by	Example
1°	Order of amino acids	Peptide bond	Similar to the order of beads on a necklace
2°	Patterns form	Hydrogen bonding	α-helix β-sheet
3°	Overall shape	Disulfide bonds Noncovalent bonds	What the protein looks like in 3D
4°	Different peptide chains associate	Noncovalent bonds	Hemoglobin has four subunits per protein

---

Chaperone proteins (or chaperonins) help fold proteins correctly.

 Ask Yourself...

1. If a protein was placed in water, where would hydrophobic amino acids be located in a folded protein and why? What about hydrophilic amino acids?
2. If a protein is embedded in the plasma membrane, where would hydrophobic amino acids be located and why? What about the hydrophilic amino acids?
3. Disrupting which level of protein structure would be the most disruptive to a protein's function?

 Antibodies (or immunoglobulins) also have quaternary protein structure.

Lipids ❗

Lipids, or fats, are the most energy-packed macromolecule. They are hydrophobic, and are found in cells in three main forms:

1. Triglycerides
2. Phospholipids
3. Cholesterol

Triglycerides ❗

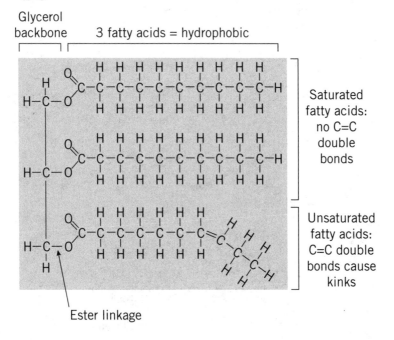

Glycerol backbone

3 fatty acids = hydrophobic

Saturated fatty acids: no C=C double bonds

Unsaturated fatty acids: C=C double bonds cause kinks

Ester linkage

Function: Energy storage, especially in fat cells (adipocytes)

Phospholipids ❗

Similar in structure to a triglyceride except:

- There are only two hydrophobic fatty acids added to glycerol
- Third position has a polar phosphate group added instead

Because phospholipids have a polar/hydrophilic end and a nonpolar/hydrophobic end, they are called **amphipathic**. This allows phospholipids to form bilayers:

- Hydrophobic fatty acids associate in the middle of the bilayer
- Hydrophilic head groups face outside

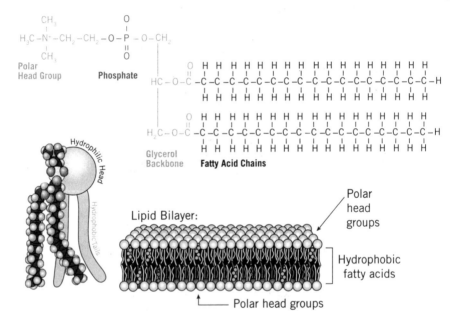

Function: Major component of the plasma membrane

Cholesterol ❗

- Four carbon rings
- Nonpolar
- Carried in the blood by high density lipoprotein (HDL) and low density lipoprotein (LDL)

Where two lines come together or end is a carbon atom.

Functions:

- Building block of steroid hormones (such as estrogen, progesterone, and testosterone)
- Optimizes membrane fluidity

Ask Yourself...

Lipids and carbohydrates are both made of oxygen, hydrogen, and carbon. Why are they so different chemically and functionally?

Nucleic Acids ❗

There are two types of nucleic acids: DNA and RNA. Both are chains (like polypeptides) but the repeating monomers are molecules called nucleotides. Nucleic acids encode biological information and this manual instructs the cell how to make proteins.

Nucleotides ❗

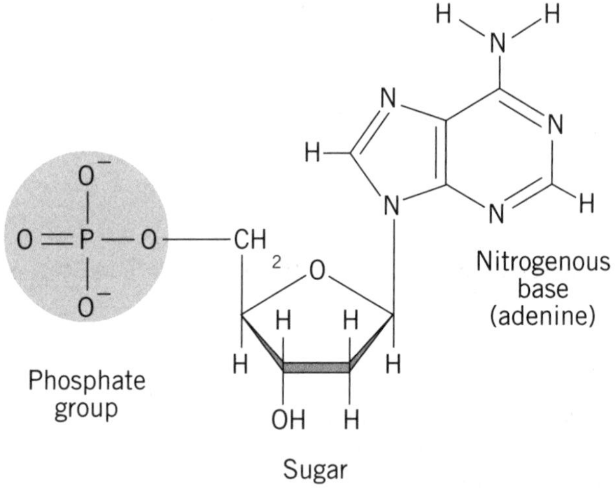

Nucleotides contain:

- Negatively charged phosphate group
- Five carbon ring or sugar: ribose or deoxyribose
- Nitrogenous base

There are five different nitrogenous bases (C, T, U, A, and G), and each is either a pyrimidine or a purine.

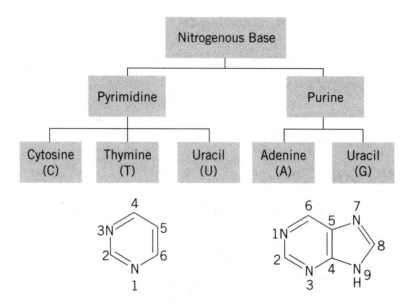

Remember!

Use "CUT the Py" to remember the pyrimidines are C, U, and T.
Use "PURe silver" to remember the purines are A and G: Ag on the periodic table is silver.

DNA ❗

Nucleotides are Connected Together

- Nucleotides are connected by sugar-phosphate phosphodiester bonds
  - One end of a nucleotide chain has a free 5' phosphate
  - Other end has a free 3' deoxyribose
- The order of nitrogenous bases functions like a code to spell out different genetic messages

DNA is Double Stranded

- DNA strands are complementary
  - Each pyrimidine hydrogen bonds to a specific purine
  - A = T
  - C ≡ G
- DNA strands are antiparallel:
  - 5' Phosphate ------------------ OH 3'
  - 3' HO ------------------ Phosphate 5'
- DNA is a right-handed double helix

Ask Yourself...

1. What would be the complementary DNA sequence to 5' – ATGCATCCGTTAGC – 3'?
2. Which is stronger, the bond between A-T or C-G? Why?

DNA vs. RNA ❗

	DNA	RNA	
Sugar	Deoxyribose	Ribose	❗
Pyrimidines	C and T	C and U	❗
Base pairing	$C \equiv G$ $A = T$	$C \equiv G$ $A = U$	❗
Stranded	Double	Single	❗
Function	Genome	Gene expression	💬
Stability	High	Low	💬
Lifespan	Long	Short	💬
Location	Nucleus Mitochondrial matrix	Nucleus Cytoplasm Ribosome	💬
Propagation	Self replicating	Transcribed from DNA	💬

Ask Yourself...

1. Why is there DNA in the mitochondrial matrix? From an evolutionary point of view, where did it come from?
2. What is the difference between ribose and deoxyribose?

Genomes

A genome is the complete set of genes or genetic material present in an organism.

Bacteria DNA   Plasmids

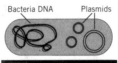

Eukaryotes	Prokaryotes	Viruses
• Many linear chromosomes	• Single circular chromosome	• Linear or circular
• Double-stranded DNA	• Double-stranded DNA	• Double-stranded or single-stranded
• Centromeres: narrow part where the two arms connect	• Can also have plasmids: smaller than chromosomes but also double-stranded DNA	• DNA or RNA
• Telomeres: chromosome ends		• Very small

 Ask Yourself...

1. How many unique chromosomes do you have in each cell?
2. What is the total number of chromosomes you have in each of your cells? How many did you inherit from your mother, and how many from your father?

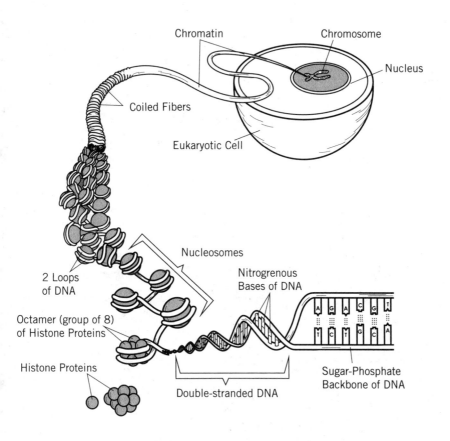

Chromatin

Chromosome

Nucleus

Coiled Fibers

Eukaryotic Cell

Nucleosomes

2 Loops
of DNA

Nitrogrenous
Bases of DNA

Octamer (group of 8)
of Histone Proteins

Histone Proteins

Sugar-Phosphate
Backbone of DNA

Double-stranded DNA

Eukaryotes have large genomes, which are densely packed to fit within the nucleus of the cell.

The DNA in a human cell is about 2.2 meters in length.
You can go to the moon and back on your DNA: if you unwound all the DNA in your body, it would be longer than a trip to the moon and back.

The human genome contains 24 different chromosomes (22 autosomes, plus two different sex chromosomes), 3.2 billion base pairs, and codes for about 21,000 genes.

Our DNA is packed so well, it only occupies about 9% of the volume of the nucleus.

Nucleosomes look like microscopic beads on a string.

CHAPTER 2

Cells

Cells are called the "basic unit of life" since all known living things
are composed of one or more cells. Living things made of a single
cell are unicellular and living things made of more than one cell are
multicellular.

Living Things ❗

All living things must have the ability to grow, reproduce, respond to stimuli, and obtain and use energy. A cell is a protective organizational system that allows these things to occur.

What Are the Different Types Cells? ❗

There are two types of cells: prokaryotic cells and eukaryotic cells. Within each of these types of cells are special subcellular components called organelles. Prokaryotic cells are generally simpler and lack most organelles. Eukaryotic cells are more complex with many interesting organelles.

Prokaryotic Cells ❗

- Simple
- Found as unicellular living things
- Found in the bacterial and the archaea domains
- No nucleus
- No membrane-bound organelles such as mitochondria, endoplasmic reticulum, or lysosome.
- Have ribosomes
- Always have cell wall

Eukaryotic Cells ❗

- Complex
- Found as unicellular living things or as part of multicellular living things
- Found in the eukarya domain, which includes fungi, protists, plants, and animals
- Many organelles
- Sometimes have cell wall

Organelles 🔊

Just like the individual parts of your body with special functions are called organs, the parts of cells with special functions are called organelles.

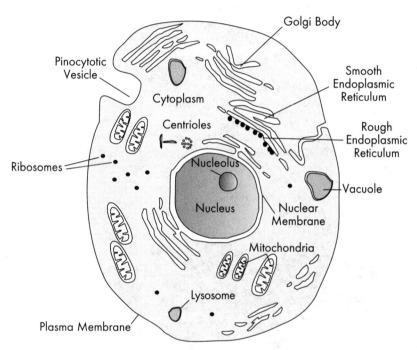

Plasma Membrane 🔊

A lipid bilayer membrane barrier separating the inside and outside of all cells and holding the cellular contents together. Membranes contain phospholipids, cholesterol, and proteins.

Fluid-Mosaic Model 💬

The fluid-mosaic model describes membranes as being flexible and a mosaic of components:

- Phospholipids: the main component
- Cholesterol: a bulk lipid that affects membrane fluidity
- Proteins: help molecules/signals get through the membrane
  - Transmembrane proteins (pass all the way through)
  - Integral membrane proteins (embedded in the membrane)
  - Peripheral membrane proteins (sit on the membrane)
- Carbohydrates: chains attached to the membrane on the extracellular side to label the cell and help with adhesion

The components of each phospholipid layer can jostle around a bit like a crowd of boats on the ocean, but the members of one layer cannot cross the hydrophobic space to join the other layer.

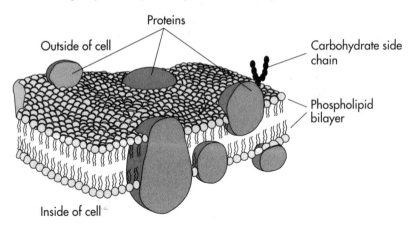

The Nucleus ❗

A membrane-bound organelle that holds and protects the DNA. It is surrounded by a double lipid-bilayer and is only found in eukaryotes. DNA replication and transcription occur here.

Ribosomes 🔔

Special protein-making factories made of rRNA. Translation occurs here. Sometimes they are located freely in the cytoplasm and sometimes they are attached to the rough ER.

Endoplasmic Reticulum 🔔

A network of canals throughout the cell that connects organelles and provides an environment similar to the extracellular space.

- Rough ER: attached to the nucleus, it is covered with ribosomes making secreted and membrane proteins
- Smooth ER: synthesizes lipids

Golgi Bodies 🔔

A series of flattened membrane sacs called cisternae that package and ship things out of the cell, and produce lysosomes.

Mitochondria 🔔

Small organelles that are the site of the Krebs cycle and the electron transport chain in eukaryotes, which lead to the production of ATP.

They have a smoother outer membrane and a squiggly inner membrane called the crista. These squiggles increase the surface area for the electron transport chain. The innermost region is the mitochondrial matrix.

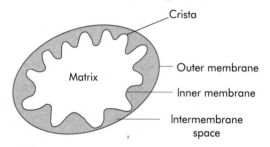

 Only the egg transfers mitochondria to offspring so you only inherit mitochondria from your mother.

Cell Wall ❗

A stiff barrier made of polysaccharides that surrounds the cell outside of the plasma membrane. It is found in plants, bacteria, and fungi.

 Ask Yourself...

What might be some downsides to having a cell wall?

Chloroplasts ❗

Specialized organelles found in algae and plants that capture energy through photosynthesis.

Inner Structures 💬

Inside the chloroplasts are membrane-bound sacs called thylakoids that are stacked in columns called grana.

Green chlorophyll molecules inside the thylakoids capture energy from sunlight and the chloroplasts use it to produce ATP and $NADPH_2$ in a series of reactions called the light-dependent reactions.

The space surrounding the thylakoid stacks is called the stroma. This is where the light-independent reactions occur by which ATP and $NADPH_2$ are used to make sugar molecules from CO_2.

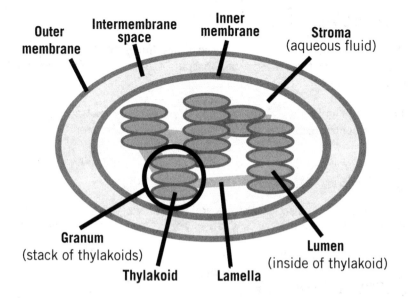

Outer membrane · **Intermembrane space** · **Inner membrane** · **Stroma** (aqueous fluid)

Granum (stack of thylakoids) · **Thylakoid** · **Lamella** · **Lumen** (inside of thylakoid)

Lysosomes

Membrane-enclosed sacs that contain hydrolytic enzymes to breakdown things within the cell.

Ask Yourself...

Where would you expect to have the highest proportion of lysosomes, an immune cell that attacks bacteria or a skin cell?

 Chlorophyll appears green because it absorbs all other wavelengths of visible light and reflects the green colored light.

Centrioles ❗

Paired cylindrical organelles that produce microtubules, commonly featured during the creation of the spindle in mitosis. They are not common in plants.

Two centrioles join with over a hundred proteins to make an organelle called a **centrosome.** This will be discussed more in Chapter 7.

Vacuoles ❗

A membrane-bound sac that stores and sometimes digests things within the cell. Plants tend to have a large vacuole that is helpful for plant stability.

Cytoskeleton ❗

A network of proteins that provides a structural framework for the cell. It is divided into three types of fibers:

- Microtubules: dynamically made of α- and β-tubulin building blocks and used for mitotic spindle, cilia, and eukaryotic flagella
- Microfilaments: dynamically made of actin building blocks and used for changing the cell shape such as during motility, cytokinesis, and endocytosis
- Intermediate filaments: a collection of many types of proteins that give a semi-permanent scaffold of cell structure

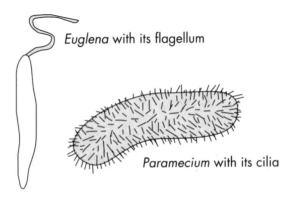

Euglena with its flagellum

Paramecium with its cilia

Differences in Cell Types

	Animals	Plants	Protists	Fungi	Eubacteria	Archaea
Eukaryotic	✔	✔	✔	✔		
Prokaryotic					✔	✔
Cell Wall		✔	Rarely	✔	✔	✔
Plasma Membrane	✔	✔	✔	✔	✔	✔
Nucleus	✔	✔	✔	✔		
Ribosomes	✔	✔	✔	✔	✔	✔
Endoplasmic Reticulum	✔	✔	✔	✔		
Golgi Bodies	✔	✔	✔	✔		
Mitochondria	✔	✔	✔	✔		
Chloroplasts		✔	Some			
Lysosomes	✔	✔	Some	✔		
Centrioles	✔	Rarely	Many	Rarely		
Vacuoles	✔ (small)	✔	✔	✔	Some	Some
Cytoskeleton	✔	✔	✔	✔	✔	✔

 Ask Yourself...

If an organism has a cell wall and a nucleus, but no chloroplasts, what type of organism is it?

Wait, what? Aren't vacuoles membrane-bound organelles? And didn't we decide bacteria don't have membrane-bound organelles?! Good catch, distinguished scientist! Only a small number of photosynthetic bacteria have vacuoles. They tend to be huge (up to 98% of cell volume) and full of either nitrates (for nitrogen storage) or gases (to help the cell control buoyancy). Either way, think of this as an exception to the rule about bacteria not having membrane-bound organelles. This is a great example of how nothing is simple in biology!

Cellular Transport: The Traffic Across Membranes !

The cell membrane plays an important role in determining what gets into and out of a cell. It is the gate that everything must pass through.

Surface Area to Volume Ratio !

Since transport is occurring at all times, the surface area must be large enough to support the molecular traffic for the cell. The ratio of the surface area to the volume of a cell must be kept as high as possible for optimal transport. Smaller cells typically have the highest ratio.

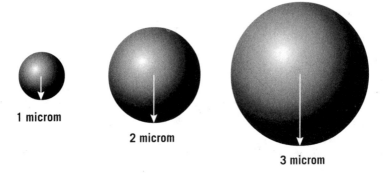

1 microm

2 microm

3 microm

Radius	1 μm	2 μm	3 μm
Surface Area ($4\pi r^2$)	12.6 μm^2	50.3 μm^2	113.1 μm^2
Volume ($4/3\pi r^3$)	4.2 μm^3	33.5 μm^3	113.1 μm^3
Surface Area to Volume Ratio	3:1	1.5:1	1:1

 Ask Yourself...

Why would the cristae of the mitochondria have lots of folds?

What Passes Through the Membrane?

Lipid membranes are a bilayer of phospholipids with a hydrophobic space in the middle. This hydrophobic sandwich only allows hydrophobic things to pass across the membrane.

Hydrophilic (charged) things need assistance from channel proteins that serve as tunnels through the membrane.

Water is well-known for its polarity. It needs a channel to cross and these water-specific channels are called aquaporins.

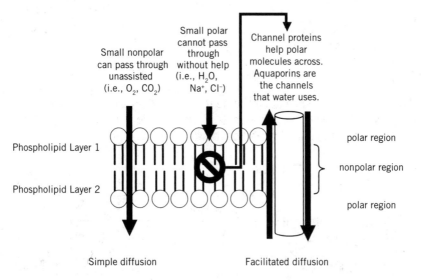

Small nonpolar can pass through unassisted (i.e., O_2, CO_2)

Small polar cannot pass through without help (i.e., H_2O, Na^+, Cl^-)

Channel proteins help polar molecules across. Aquaporins are the channels that water uses.

Phospholipid Layer 1

Phospholipid Layer 2

polar region

nonpolar region

polar region

Simple diffusion

Facilitated diffusion

Diffusion and Osmosis �︎

Molecules naturally move from areas where they are the most concentrated to areas where they are the least concentrated. This is the process of diffusion.

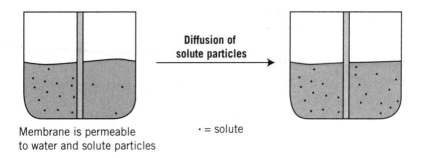

Diffusion of solute particles

Membrane is permeable to water and solute particles

• = solute

Osmosis is the specific movement of water or another solvent from where it is most concentrated (a dilute solution) to where it is least concentrated (a concentrated solution).

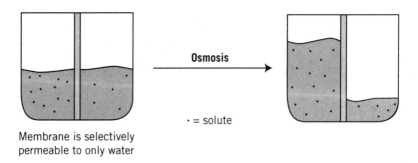

Osmosis

Membrane is selectively permeable to only water

• = solute

? Ask Yourself...

After osmosis (in the beaker on the right), which side has the higher concentration of water?

Passive Transport 🔊

Passive transport does not require the input of energy, and molecules diffuse according to their concentration gradient.

Simple Diffusion 🔊

The passive movement of small hydrophobic molecules as they pass through the membrane without assistance. O_2 and CO_2 move by simple diffusion.

Facilitated Diffusion 🔊

The diffusion of small hydrophilic molecules across the membrane with the help of a channel protein that allows them to cross the hydrophobic space.

 Ask Yourself...

Would a chloride ion travel by simple or facilitated diffusion?

Active Transport 🔊

Active transport requires the input of energy to pump molecules against their concentration gradient.

Primary Active Transport 🔊

Transport that relies on the catabolism of energy molecules such as ATP to power the active pumping of molecules against their concentration gradient. The Na^+/K^+ pump is an example of primary active transport.

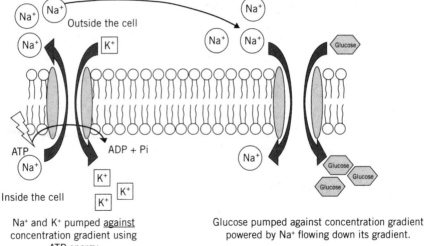

Na⁺ and K⁺ pumped against concentration gradient using ATP energy	Glucose pumped against concentration gradient powered by Na⁺ flowing down its gradient.
PRIMARY ACTIVE TRANSPORT	**SECONDARY ACTIVE TRANSPORT**

Secondary Active Transport ❶

Transport that relies on a gradient of molecules moving by passive transport to power the active pumping of other molecules against their concentration gradient. The sodium glucose co-transporter is an example of secondary active transport.

 Ask Yourself...

Where does the energy required to push glucose across its concentration gradient come from in secondary active transport?

Endocytosis and Exocytosis !

Large-scale movement of things into the cell occurs by endocytosis: the membrane envelopes content outside the cell and "ingests" it. The opposite process occurs during exocytosis, when the membrane retracts and expels content out of the cell.

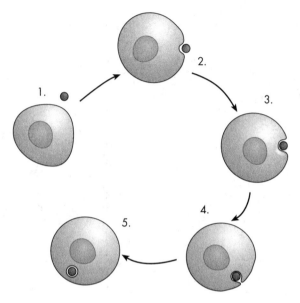

 Sometimes viruses and bacteria sneak in through endocytosis.

Cell Junctions

- Desmosomes: Typical attachments between cells. Some things can squeeze between.
- Gap junctions: Special tunnels between cells giving a direct connection of the cytoplasm.
- Tight junctions: Impenetrable barriers preventing anything from slipping between the cells.

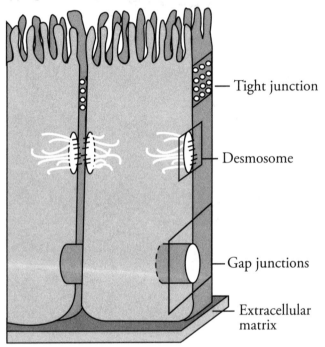

Tight junction

Desmosome

Gap junctions

Extracellular matrix

Remember...

- Desmosomes function like buttons and tight junctions function like zippers.
- Gap junctions function like bridges or tunnels between buildings.
- Tight junctions are impermeable, like the grout between tiles in your bathroom or kitchen.

 Tight junctions are found in the intestine. If they break down it can lead to severe diarrhea.

Cell Communication ❗

Cells are capable of sending and receiving signals. Signals can be stimulatory or inhibitory. This messaging is called signal transduction.

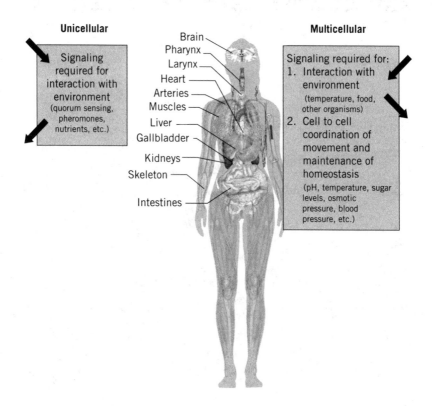

Unicellular

Signaling required for interaction with environment (quorum sensing, pheromones, nutrients, etc.)

Brain
Pharynx
Larynx
Heart
Arteries
Muscles
Liver
Gallbladder
Kidneys
Skeleton
Intestines

Multicellular

Signaling required for:
1. Interaction with environment (temperature, food, other organisms)
2. Cell to cell coordination of movement and maintenance of homeostasis (pH, temperature, sugar levels, osmotic pressure, blood pressure, etc.)

Ask Yourself...

What would happen if different systems in the body could not communicate with each other?

1. Sending the Signal ❗

Signals can be sent several ways: through direct contact, sending signals over short distances, or sending signals over long distances.

Type of Signal	Examples	
Direct contact	Immune Cells, Plasmodesmata	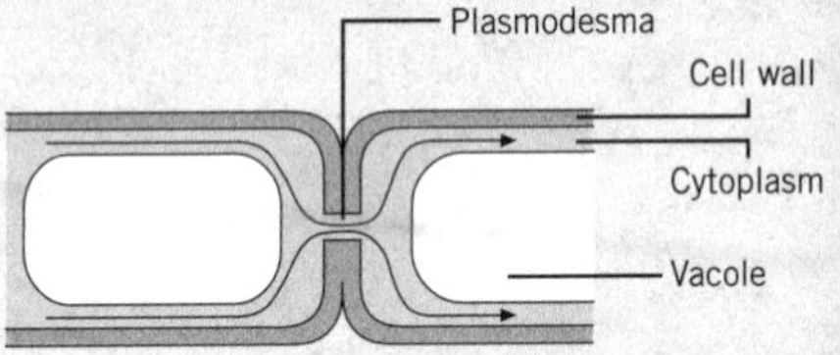
Signals sent over short distances	Neurotransmitters, morphogens in embryonic development, quorum sensing in bacteria	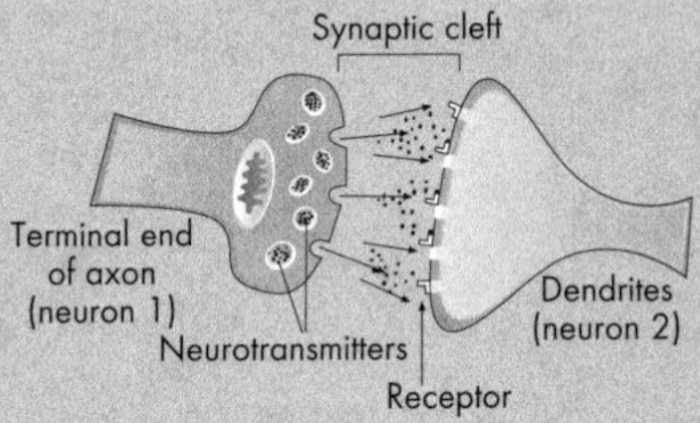
Signals sent over long distances	Hormones	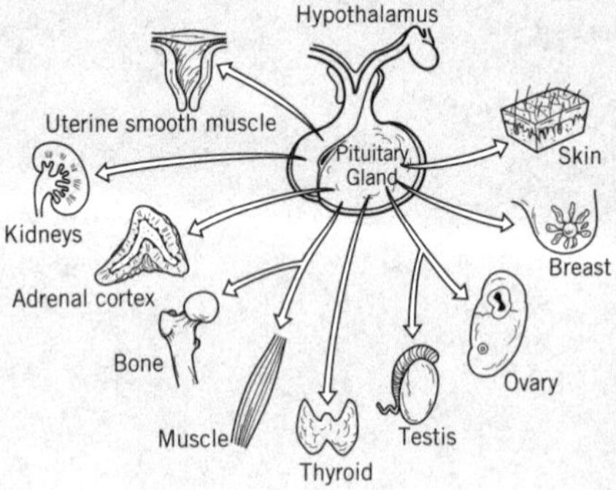

2. Receiving the Signal 🛑

When a signaling molecule arrives at a cell, it must either enter the cell and bind to an intracellular receptor or bind to an extracellular receptor on the outside. Either way, the signaling molecule is called a ligand and the molecule it binds to is called the receptor.

The binding between a ligand and a receptor is very specific and the two parts must have a coordinating shape and charge. Once binding has occurred, this often induces a conformational change in the receptor that triggers a response.

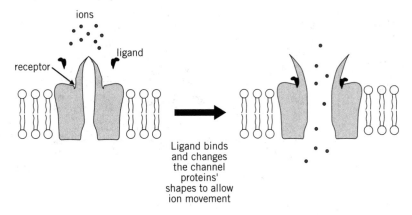

Ligand binds and changes the channel proteins' shapes to allow ion movement

? Ask Yourself...

What would happen if a molecule blocked the ligand from binding to the receptor?

Cells

3. Convert to Cellular Response 🛈

The cellular response varies depending on the signal that was received. As the message is passed, there is often sequential phosphorylation of several molecules in a chain. Often, the response is also amplified and many changes in the cell can occur from receiving a single signal.

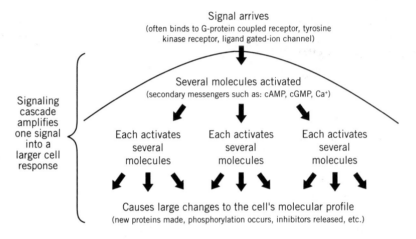

Signal arrives
(often binds to G-protein coupled receptor, tyrosine kinase receptor, ligand gated-ion channel)

Several molecules activated
(secondary messengers such as: cAMP, cGMP, Ca⁺)

Signaling cascade amplifies one signal into a larger cell response

Each activates several molecules

Each activates several molecules

Each activates several molecules

Causes large changes to the cell's molecular profile
(new proteins made, phosphorylation occurs, inhibitors released, etc.)

CHAPTER 3

Cellular Energetics

Life depends on a large number of precise reactions that occur within each cell. Enzymes help the cell perform these reactions by making them more doable. Many of these reactions are powered by ATP, which is made in cell respiration. Carbohydrates can be broken down to CO_2 in a process called oxidation. Because the process releases large amounts of energy, carbohydrates generally serve as the principal energy source for cellular metabolism. Cell respiration allows the cell to transfer the energy in carbohydrates to ATP.

Bioenergetics ❗

The study of how energy flows through living organisms.

The total amount of energy in the universe remains constant, but energy can change from one form to another.

This type of energy transformation occurs in many of the processes that take place in living things, allowing energy to flow through processes and cells.

All living organisms require a source of energy, and all living organisms use the same metabolic pathways to power life.

Organisms use energy to increase or maintain order.

Chemical bonds are the source of all energy for life

- All energy on the Earth comes from the sun
- Photosynthetic organisms use solar energy to turn CO_2 and water into carbohydrates
  - This also releases O_2
- Biological macromolecules (especially carbohydrates and fats) store energy in their chemical bonds
- Cells use respiration to transfer energy from the chemical bonds of macromolecules to the chemical bonds of smaller molecules such as ATP, GTP, NADH, NADPH or $FADH_2$
  - This releases CO_2 and H_2O
- These small, high energy molecules power all the reactions and processes the cell needs to do

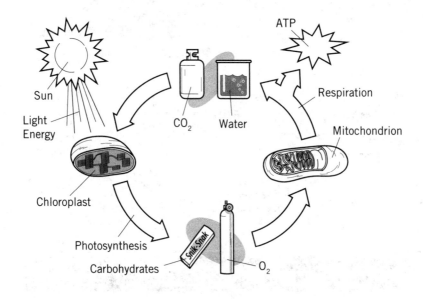

 Ask Yourself...

If all living organisms perform the same metabolic pathways, when did these pathways first evolve?

 Photosynthesis can be performed by all plants, some bacteria, and some protists.

Thermodynamics ❗

The study of energy transformations that occur in matter. Energy flow and metabolism in organisms are subject to the laws of thermodynamics.

Law	Meaning	Application
First law of thermodynamics	Energy cannot be created or destroyed. The total amount of energy in the universe is constant.	Energy can't come from nowhere. Reactions and cells must take energy from one thing to power something else.
Second law of thermodynamics	Energy transfer leads to less organization.	The universe naturally tends toward disorder, randomness, or increasing entropy (ΔS).

💬

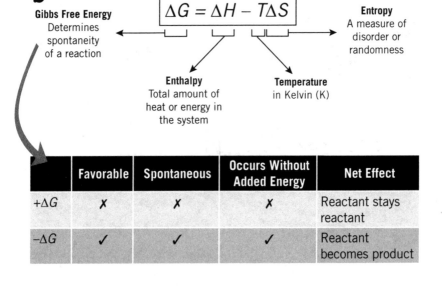

Gibbs Free Energy
Determines spontaneity of a reaction

$$\Delta G = \Delta H - T\Delta S$$

Entropy
A measure of disorder or randomness

Enthalpy
Total amount of heat or energy in the system

Temperature
in Kelvin (K)

	Favorable	Spontaneous	Occurs Without Added Energy	Net Effect
$+\Delta G$	✗	✗	✗	Reactant stays reactant
$-\Delta G$	✓	✓	✓	Reactant becomes product

Energy in Reactions ❗

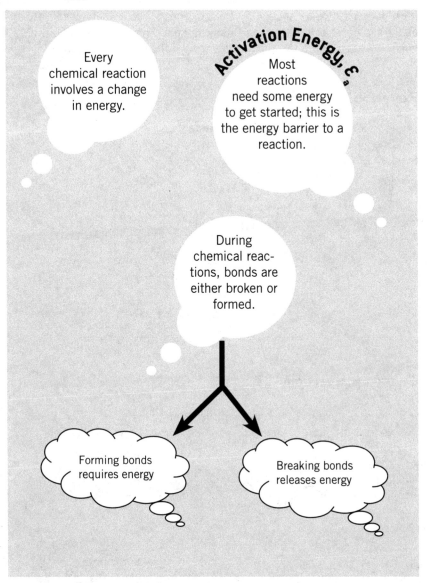

Every chemical reaction involves a change in energy.

Activation Energy, E_a

Most reactions need some energy to get started; this is the energy barrier to a reaction.

During chemical reactions, bonds are either broken or formed.

Forming bonds requires energy

Breaking bonds releases energy

 Ask Yourself...

Is cleaning your room associated with an increase or decrease in ΔS?

Types of Reactions ❗

Most reactions need an energy investment to occur, but overall are either exergonic or endergonic. Catalysts help decrease the amount of energy needed at the start, but don't change the overall reaction.

Exergonic Reactions ❗

- Products have less energy than reactants
- Energy is given off
- Favorable

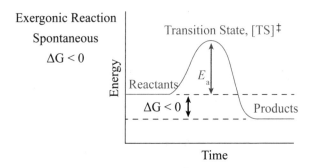

Endergonic Reactions ❗

- Products have more energy than reactants
- Require an input of energy
- Unfavorable

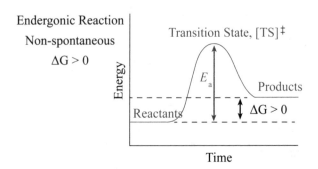

Catalyst ❗

- Lowers the activation energy barrier (E_a) of a reaction
- Makes it easier to overcome the energy barrier of the reaction
- Speeds up the rate of a reaction
- Doesn't change or affect the reaction itself
- Not used up

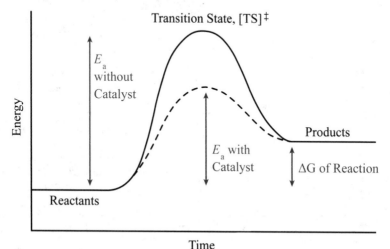

Ask Yourself...

Do all exergonic reactions release heat? Do all endergonic reactions absorb heat?

Enzymes !

Biological catalysts that increase rate of reactions by lowering the activation energy barrier of the reaction.

Enzymes are:

- Highly specific
- Usually proteins

Enzymes do:

- Control reactions essential for life
- Act on reactants called "substrates"
- Form temporary enzyme-substrate complexes
- Remain unaffected by the reaction

Enzymes don't:

- Change the reaction
- Make reactions occur that would otherwise not occur at all

Most enzymes are proteins, but RNA enzymes have also been found. Ribozymes are RNA molecules that are enzymes; they help with splicing and protein synthesis in the ribosome.

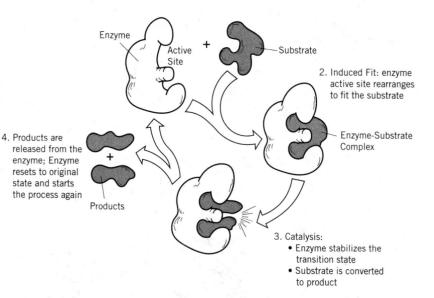

1. Substrate binds enzyme at the active site

Enzyme

Active Site

+

Substrate

2. Induced Fit: enzyme active site rearranges to fit the substrate

Enzyme-Substrate Complex

4. Products are released from the enzyme; Enzyme resets to original state and starts the process again

+

Products

3. Catalysis:
- Enzyme stabilizes the transition state
- Substrate is converted to product

Remember...

The most common way to name enzymes is to add the suffix -ase onto the name of the substrate. For example, peroxidase breaks down peroxide. Telomerase produces telomeres. Proteases break down proteins.

Ask Yourself...

The following enzymes break biochemical bonds. What molecules do you think they degrade? Maltase, DNase, lactase, RNase, sucrase, peptidase.

Binding Partners 💬

Some enzymes work alone, while others must bind to another molecule to function properly. Here are the options:

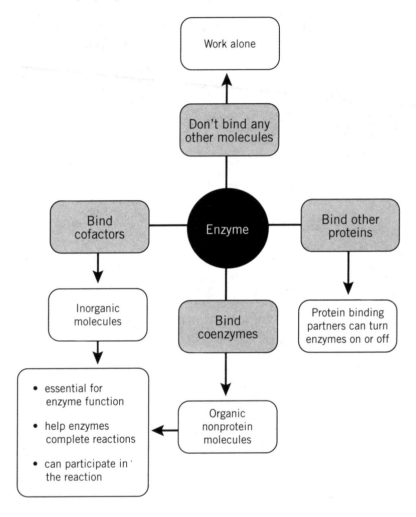

Enzyme Regulation 💬

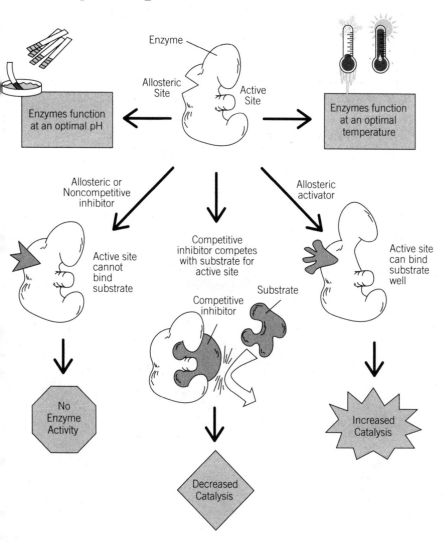

Enzyme

Allosteric Site

Active Site

Enzymes function at an optimal pH

Enzymes function at an optimal temperature

Allosteric or Noncompetitive inhibitor

Allosteric activator

Active site cannot bind substrate

Competitive inhibitor competes with substrate for active site

Active site can bind substrate well

Competitive inhibitor

Substrate

No Enzyme Activity

Decreased Catalysis

Increased Catalysis

Remember...

Human cells are usually at a pH of about 7.4 and a temperature of around 37°C / 98.6°F. This means most of your enzymes function best at these conditions.

Ask Yourself...

1. Hydrolase enzymes in the lysosome help your cells degrade old organelles. How well do these enzymes function at pH = 7.4?
2. What pH results in optimal activity of the enzymes found in your stomach?

Q_{10} is a measure of temperature sensitivity of a physiological process or enzymatic reaction rate due to a change in temperature. The more temperature-dependent a reaction is, the higher Q_{10} will be.

Ask Yourself...

1. If $Q_{10} = 1$, is a reaction temperature dependent or temperature independent?
2. Most physiological reactions have a Q_{10} between 2 and 3. This means their reaction rate changes by _____ when the temperature changes by 10K?

Regulating Enzymes in Biochemical Pathways ❗

Most enzymes are part of a larger pathway or system in the cell. This means they are often regulated as part of a larger biochemical pathway.

Negative Feedback ❌
Maintains dynamic homeostasis

Positive Feedback ✔
Amplifies responses/processes

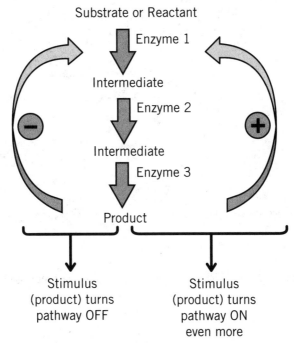

Substrate or Reactant

Enzyme 1

Intermediate

Enzyme 2

Intermediate

Enzyme 3

Product

Stimulus (product) turns pathway OFF

Stimulus (product) turns pathway ON even more

For examples of positive and negative feedback, go online and check out the ASAP Biology Supplement file posted there!

Powering Biological Reactions: ATP !

Adenosine triphosphate (or ATP) contains phosphate bonds that store potential energy. This means that ATP hydrolysis is exergonic.

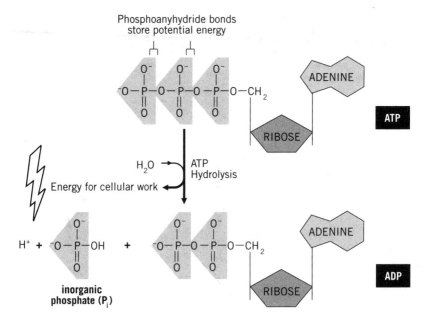

The cell can use ATP to power other expensive (endergonic) reactions, such as protein synthesis.

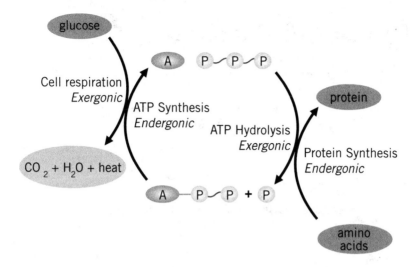

Sources of ATP:

1. Light reactions of photosynthesis
2. Cell respiration

 Ask Yourself...

1. Is ATP the only molecule in your cell that contains high energy phosphate bonds?
2. Can ATP be used as a building block in DNA, RNA, or both?

 There are lots of everyday examples of this reaction coupling. For example, studying can be a bit of a pain, but getting a good mark on something feels great. Staying up late can be tiring, but if you're having fun with your friends, it is worth it.

Aerobic Cell Respiration

Occurs in the presence of oxygen.

Overall Reaction for Cell Respiration:

Glucose $(C_6H_{12}O_6) + 6O_2 \rightarrow 6CO_2 + 6H_2O + 30ATP$

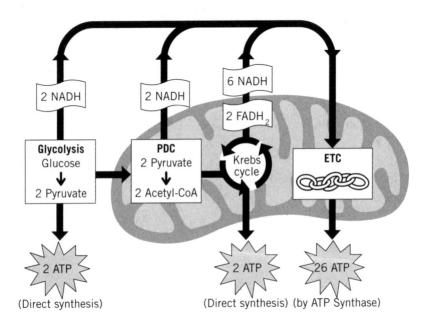

2 NADH

2 NADH

6 NADH

2 FADH$_2$

Glycolysis
Glucose
↓
2 Pyruvate

PDC
2 Pyruvate
↓
2 Acetyl-CoA

Krebs
cycle

ETC

2 ATP
(Direct synthesis)

2 ATP
(Direct synthesis)

26 ATP
(by ATP Synthase)

Remember...

Cell respiration is a series of redox reactions. Glucose is being oxidized to make energy (or ATP).

---

1. The pyruvate dehydrogenase complex (or PDC) links glycolysis and the Krebs cycle.
2. Pyruvic acid is an acid. In your cells, it is usually loses a proton, so is called pyruvate instead.
3. The electron transport chain is usually abbreviated to ETC.

Carbon Flow

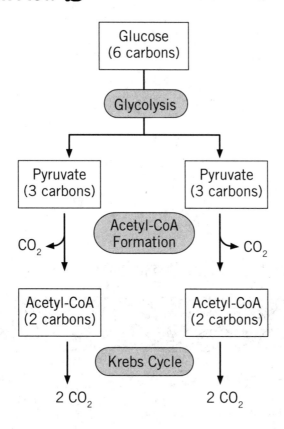

Glucose
(6 carbons)

Glycolysis

Pyruvate
(3 carbons)

Pyruvate
(3 carbons)

CO_2

Acetyl-CoA
Formation

CO_2

Acetyl-CoA
(2 carbons)

Acetyl-CoA
(2 carbons)

Krebs Cycle

$2\ CO_2$

$2\ CO_2$

 Remember...

Glucose is a monosaccharide, or carbohydrate.

The molecules made in glycolysis can be shuttled out to other metabolic pathways your cell is running. There are usually dozens of these going at the same time in a healthy cell. They help your cells make biomolecules and organelles, and perform the basic functions of life.

Energy Flow !

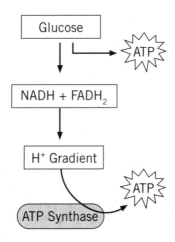

Locations and Net Output !

Process	Location	Net Output	
		Carbon Products (per glucose)	Other Products (per glucose)
Glycolysis	Cytoplasm	2 pyruvate	2 NADH 2 ATP
Formation of Acetyl-CoA by the PDC	Mitochondrial matrix	2 acetyl-CoA 2 CO_2	2 NADH
Krebs or Citric Acid Cycle	Mitochondrial matrix	4 CO_2	6 NADH 2 $FADH_2$ 2 ATP
ETC and Chemiosmosis	Mitochondrial inner membrane	None	26 ATP
Total		**6 CO_2**	**30 ATP**

(We haven't bothered with O_2 and H_2O in this table.)

 Remember... !

Locations are important! Make sure you remember these!

 NADH and $FADH_2$ are reduced molecules. They move a few different things around the cell: hydrogen atoms, electrons and potential energy.

ATP in Glycolysis

- Glucose has to be activated (by phosphorylation) at the beginning of glycolysis

- That's why glycolysis uses 2 ATP molecules

- Glycolysis also generates 4 ATP later on

- Net = 2 ATP per glucose in glycolysis

Remember...

...there are three types of ETCs:

1. Eukaryotic inner mitochondrial membrane—cell respiration—terminal electron acceptor: O_2

2. Prokaryotic plasma membrane—cell respiration—terminal electron acceptor: O_2

3. Eukaryotic chloroplast thylakoid membrane—photosynthesis—terminal electron acceptor: $NADP^+$

Chemiosmosis

- Movement of ions across a membrane, down their electrochemical gradient

- Example: hydrogen ions cross a membrane to make ATP in both cellular respiration and photosynthesis

Oxidative Phosphorylation

- A metabolic pathway where cells use enzymes to oxidize nutrients

- This releases energy

- The energy is used to make ATP

ATP Yield

- Each cytoplasmic NADH = 2.5 ATP

- Each mitochondrial NADH = 1.5 ATP

- Each $FADH_2$ = 1.5 ATP

- Each glucose = 30 ATP

Electron Transport Chain

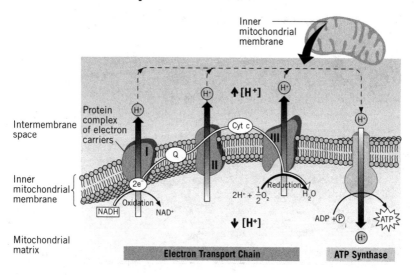

Inner mitochondrial membrane

Intermembrane space

Protein complex of electron carriers

Inner mitochondrial membrane

Mitochondrial matrix

$\uparrow$ [H$^+$]

Cyt c

Q

I

II

III

2e

Oxidation

NADH

NAD$^+$

$2H^+ + \frac{1}{2}O_2$

Reduction

H_2O

$\downarrow$ [H$^+$]

ADP + P$_i$

ATP

Electron Transport Chain

ATP Synthase

Ask Yourself...

Bacterial cells don't have mitochondria. Where do they perform the Krebs cycle? What about the electron transport chain?

All the numbers and locations we've talked about in this chapter are for eukaryotic cells.

Prokaryotic cells perform cell respiration in the cytoplasm and plasma membrane, and generate 32 ATP per glucose.

 ETC is performed by a series of molecules in a membrane doing redox reactions. Some of the carrier molecules in the electron transport chain are iron-containing carriers called cytochromes.

Cellular Energetics

Anaerobic Cell Respiration

Occurs in the absence of oxygen.

Why? 💬

- Anaerobic means "in the absence of oxygen"
- No O_2 means no ETC... means no Krebs cycle... means cells have to survive on only glycolysis
- Without the ETC, cells run out of NAD^+, and glycolysis needs NAD^+
- Fermentation regenerates NAD^+ so glycolysis can keep running without the ETC

How? ❗

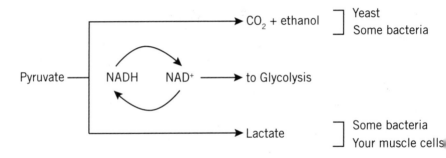

Remember...

Anaerobic respiration is also known as fermentation.

Lactic acid is an acid. In your cells, it is usually loses a proton, so is called lactate instead. When you're sore after working out, it is due to lactic acid build up in your skeletal muscles.

CHAPTER 4

Molecular Genetics

DNA is called the blueprint of life because it contains the instructions for making living things. By controlling protein synthesis, it tells our cells how and when to grow, develop, survive and reproduce. Almost every cell in your body contains DNA, and in eukaryotes like us, DNA is organized into linear chromosomes and stored within the nucleus.

DNA = Heredity

It took a lot of work to figure out the structure of DNA, and to prove it is the carrier of genetic information and heritable traits.

 Remember!

Watson and Crick first reported the structure of DNA. This won them the 1962 Nobel prize.

 For more information on how scientists determined the importance of DNA, go online and check out the ASAP Biology Supplement file posted there!

Why is DNA Important? ❗

DNA instructs our cells how to live, survive, and adapt. The central dogma of molecular biology explains the flow of genetic information in a biological system, from DNA to RNA to protein.

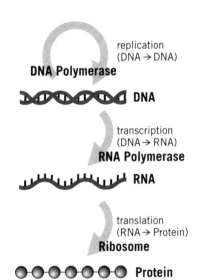

replication
(DNA → DNA)
DNA Polymerase

DNA

transcription
(DNA → RNA)
RNA Polymerase

RNA

translation
(RNA → Protein)
Ribosome

Protein

 Ask Yourself...

Is the central dogma always a one way flow of information? Does it ever occur backwards?

DNA Replication

Because DNA is the blueprint of life and contains our genetic and hereditary information, it must be passed from one generation to the next during reproduction, and from mother cells to daughter cells during cell division.

Important Key Points about DNA Replication ❗

- Occurs during the S phase of the cell cycle
- Must be complete before cell division can occur
- Semiconservative
- Requires a template and primers
- Main enzyme: DNA polymerase
- New DNA is built 5' to 3'
- DNA template is read 3' to 5'

DNA replication is semiconservative.

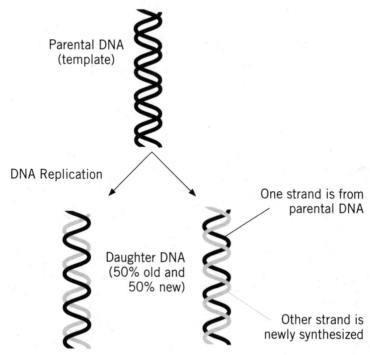

Parental DNA
(template)

DNA Replication

Daughter DNA
(50% old and
50% new)

One strand is from
parental DNA

Other strand is
newly synthesized

DNA replication starts at the origin of replication (ORI) and occurs bidirectionally in a replication bubble

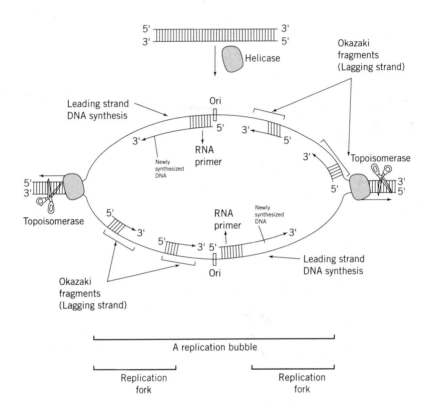

A replication bubble

Replication fork

Replication fork

Remember!

The <u>lead</u>ing strand is made continuously and <u>leads</u> into the replication fork. The lagging strand is made discontinuously.

Ask Yourself...

Does DNA polymerase add nucleotides onto the 3' or 5' end of DNA?

DNA Replication Requires Many Enzymes ❗

	Enzyme	Function

1	Helicase	• Unwinds double-stranded DNA, using ATP
		  ○ Breaks hydrogen bonds between the two strands
		  ○ Starts at the ORI
		• Creates and widens the replication bubble

2	Topoisomerase	• Snips one or both strands of template DNA • Relieves tension created by helicase • Prevents tangling or breaking of the template DNA
3	Primase	• RNA polymerase • Lays down RNA primer

4	DNA polymerase	• Builds new DNA 5' to 3' by reading template DNA 3' to 5'
		  ○ Starts at the primer
		  ○ New DNA is complementary to template DNA
		• Replaces RNA primer with DNA
		• Can repair mistakes in DNA base pairing

5	DNA ligase	• Connects fragments of DNA in the replication bubble, including:
		  ○ Okazaki fragments to each other
		  ○ Leading strand to lagging strand

 Ask Yourself...

Is it easier for helicase to separate an A/T rich region of DNA, or a C/G right region of DNA? Why?

Prokaryotes vs. Eukaryotes

Eukaryotes have linear chromosomes and form replication bubbles.

Replication Bubbles

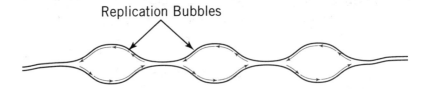

Prokaryotes have circular chromosomes and perform theta replication.

Origin of Replication

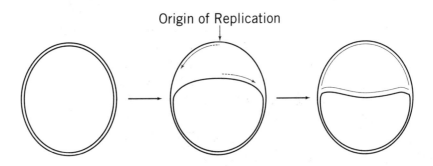

Protein Synthesis

Our cells use three steps to make proteins from DNA. **Transcription** turns DNA in the nucleus into an RNA message. This message is edited and capped, then moved to the cytoplasm. There, the ribosome reads the RNA message and **translates** the information into a protein.

Types of RNA

Three types of RNA allow cells to turn the information coded in DNA into proteins. They are all made from a DNA template via transcription.

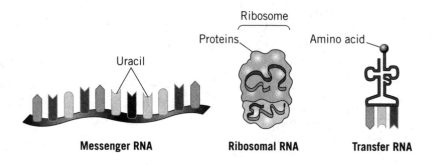

Messenger RNA   Ribosomal RNA   Transfer RNA

mRNA	rRNA	tRNA
Read by ribosome, as a template for making proteins	Major component of ribosome	Binds amino acids and brings them to the ribosome

Remember!

RNA contains the nitrogenous base uracil, instead of thymine.

 Eukaryotic cells have three additional types of RNA that you will learn about later in this chapter and in your online resources. They are hnRNA, miRNA and siRNA.

Transcription ❗

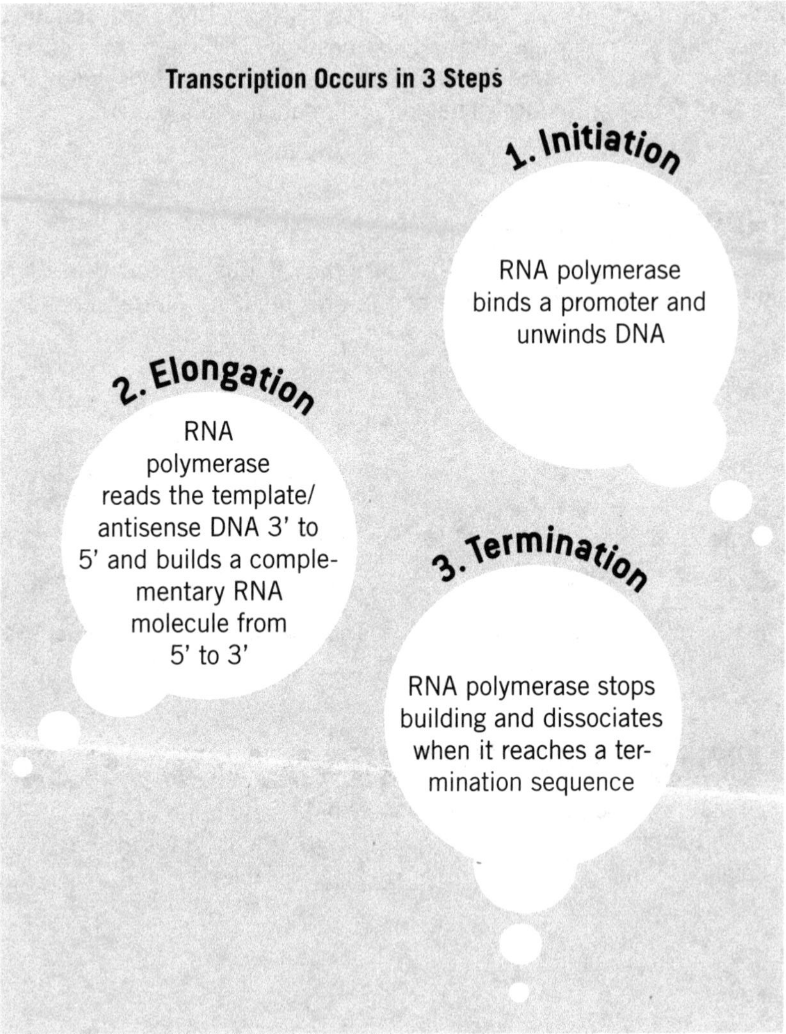

Transcription Occurs in 3 Steps

1. Initiation

RNA polymerase binds a promoter and unwinds DNA

2. Elongation

RNA polymerase reads the template/antisense DNA 3' to 5' and builds a complementary RNA molecule from 5' to 3'

3. Termination

RNA polymerase stops building and dissociates when it reaches a termination sequence

Remember!

Helicase unwinds double-stranded DNA for DNA replication. RNA polymerase unwinds double-stranded DNA for transcription.

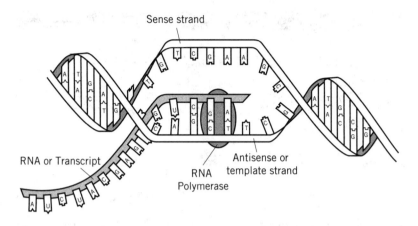

Sense strand

RNA or Transcript

Antisense or
template strand

RNA
Polymerase

In eukaryotes, transcription makes an immature RNA molecule called
hnRNA (heteronuclear RNA). This molecule is processed to become a
mature mRNA. In prokaryotes, transcription and translation happen at the
same time, so the RNA molecule that is made must be mature and ready to
be translated right away.

 Remember!

- The sense DNA strand is the same sequence (or sense) as the RNA
  being made (except for the T/U switch).

- The antisense strand of DNA is complementary to both the sense
  strand of DNA and the RNA it codes for.

 Ask Yourself...

If an antisense strand has the sequence 5' – ATCGATCGATTCGATTCG – 3',
what is the sequence of the sense strand? RNA?

DNA Replication vs. Transcription

	Replication	Transcription
Where does it start?	ORI	Promoter
What is copied?	Both DNA strands	One DNA strand at a time
What is made?	DNA	RNA
What is the main enzyme?	DNA polymerase	RNA polymerase

RNA Processing occurs in 4 steps:

1. **Splicing:** Introns are removed from hnRNA and exons are ligated together.
2. **Add a Cap:** A methylated guanine nucleotide (GTP) is added to the 5' end of the RNA
3. **Add a Tail:** A string of adenine nucleotides (or a poly-A tail) is added to the 3' end of the RNA
4. **Transport:** Finally, the mature mRNA is moved from the nucleus to the cytoplasm

To see what RNA processing looks like, go online and check out the ASAP Biology Supplement file posted there!

Remember!

Exons are expressed and introns go in the trash.

Splicing is performed by a large complex called the spliceosome. It contains over 100 proteins and some RNA molecules.

Adding a cap and tail to the RNA helps stabilize it.

The Genetic Code

- The Genetic Code translates the language of nucleic acids (A, T, G, C) into a sequence of amino acids that are part of a peptide chain
- There are 64 codons and each has 3 nucleotides
- Each codon specifies one amino acid or a stop codon
- Since there are only 20 amino acids, most can be coded for by more than one codon

 To see the genetic code and a few examples, go online and check out the ASAP Biology Supplement file posted there!

Ask Yourself...

1. Why does the Genetic Code contain uracil (U) and why is thymine (T) absent?
2. AUG is the start codon, and initiates translation. What is the complementary anticodon? What is the DNA sequence (sense and antisense strand) that code for this triplet?

Remember!

The three stop codons are 3 ways to say "stop": U Are Away, U Are Gone, U Go Away.

Major features of the genetic code are shared by all modern living systems.

 Two or more codons coding for the same amino acid are known as synonyms. Because it has such synonyms, the genetic code is said to be "degenerate."

Translation ❗

Translation Occurs in 3 Steps

1. Initiation

At an AUG codon close to the 5' end of the mRNA, a ribosome assembles and a tRNA delivers methionine to the P site.

2. Elongation

- tRNA-amino acid arrives at the A site
- Peptide bond is formed between the amino acid in the P site and the amino acid in the A site
- Ribosome shifts over three nucleotides (one codon) to free up the A site

3. Termination

The ribosome encounters a stop codon and dissociates, freeing the peptide chain.

Translation 💬

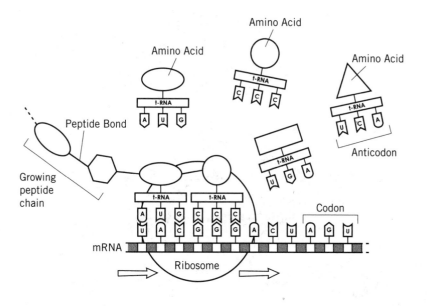

Ribosomes are made of rRNA and proteins. Each has a small subunit and a large subunit. The large subunit has three sites: E, P and A.

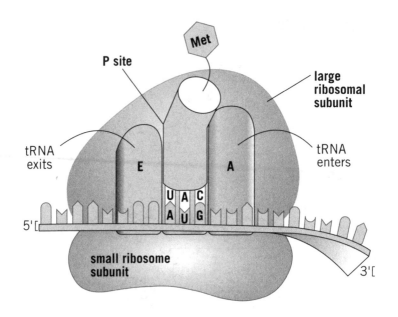

Remember! .

Amino acids <u>a</u>rrive at the <u>A</u> site, the <u>p</u>eptide chain winds out of the <u>P</u> site and the tRNA <u>e</u>xits the <u>E</u> site.

For a summary of how eukaryotes and prokaryotes differ in molecular biology, go online and check out the ASAP Biology Supplement file posted there!

Mutations 🛑

People often think mutations are a bad thing (and lots are!), but many mutations are neutral (they have no effect), or positive (they confer some sort of advantage). In fact, mutations are the primary source of genetic variability.

Where do Mutations Come From?

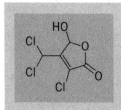

Mutagens, Chemicals, Drugs

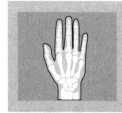

Light
- Ionizing radiation (X-rays, α particles, γ rays)
- UV

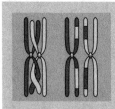

Errors in Normal Cell Processes
- DNA replication (especially DNA polymerase)
- DNA repair
- Recombination in meiosis
- Chromosome segregation in mitosis and meiosis

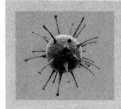

Biological Agents
- Lysogenic viruses

 For more information on how we deal with mutations, go online and check out the ASAP Biology Supplement file posted there!

 Ask Yourself...

Are mutations more serious when they occur in DNA or RNA?

❗ There are 2 Types of Mutations:

1. Base substitutions
2. Gene rearrangements

💬 Base substitutions are also called point mutations. They are when one nucleotide is changed to another. For example: AAG → ACG. There are three kinds of point mutations:

Nonsense	Missense	Silent
Codon is changed to a stop codon	Codon is changed to a different codon	Codon is changed to a different codon
Peptide chain is shorter (or truncated)	One amino acid is switched to another	No change in amino acid (because the Genetic Code is degenerate)
Example: AGA → UGA Arg → Stop	Example: CUU → CCU Leu → Pro	Example: GGU → GGG Gly → Gly

 Generally, silent mutations do not significantly alter phenotype. They can also occur in non-coding regions of DNA (outside of genes or within introns).

Gene Rearrangements 💬

Mutation	Definition	Example
Insertion	Add 1 to thousands of bases	AAGTTC → AACGTTC
Deletion	Delete 1 to thousands of bases	AAGTTC → AAGTC
Inversion	Chromosome section is flipped	
Amplification Or Duplication	Chromosome section is duplicated	
Translocation	Chromosome segment is swapped with another (nonhomologous) chromosome	

Inversion figure:

a b c d e f g h i
Chromosome A

Mutant Chromosome A
a b c f e d g h i

Amplification Or Duplication figure:

Gene 1
Gene 2
Gene 3

Gene duplication, increase copy number →

Translocation figure:

Before translocation — Normal chromosome 20, Normal chromosome 4

After translocation — Mutant chromosome 20, with some of chromosome 4; Mutant chromosome 4, with some of chromosome 20

Ask Yourself...

Does an insertion that adds 6 nucleotides to a gene cause a frameshift?

Insertions and deletions are often called "indels" and grouped together as frameshift mutations. Because nucleotides are read in three base codons when the cell is making a protein, indels can shift the reading frame of translation.

Molecular Genetics

Controlling Gene Expression

Genomes contain lots of genes; for example, your genome contains about 21,000 genes. However, not all genes are expressed in every cell. Each cell type (in you and in all organisms) expresses many genes but keeps other silent. This means cells need a way to turn gene expression on and off, and this can be coupled to events happening in the cell, signals from nearby cells, or stimuli from the external environment. The genes expressed in a particular cell determine what that cell can do.

Structure of a Gene 💬

All genes contain three things:

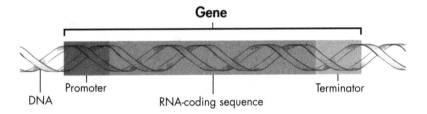

There can also be additional sequences nearby that help control the level of transcription for a certain gene.

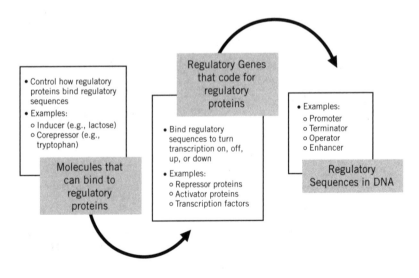

In a given cell, some genes are always expressed and others are never expressed. Some genes need to be on in certain situations and off in others—these genes have positive and negative control mechanisms to help regulate their expression. For example:

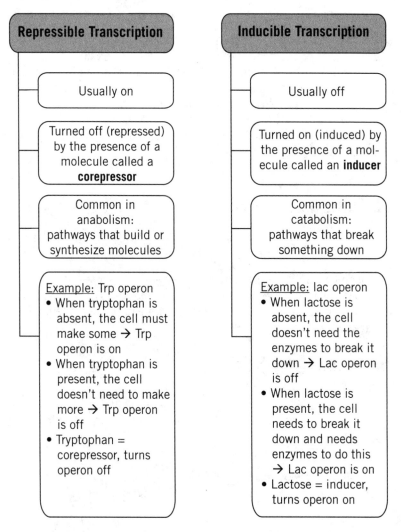

Repressible Transcription	Inducible Transcription
Usually on	Usually off
Turned off (repressed) by the presence of a molecule called a **corepressor**	Turned on (induced) by the presence of a molecule called an **inducer**
Common in anabolism: pathways that build or synthesize molecules	Common in catabolism: pathways that break something down
Example: Trp operon • When tryptophan is absent, the cell must make some → Trp operon is on • When tryptophan is present, the cell doesn't need to make more → Trp operon is off • Tryptophan = corepressor, turns operon off	Example: lac operon • When lactose is absent, the cell doesn't need the enzymes to break it down → Lac operon is off • When lactose is present, the cell needs to break it down and needs enzymes to do this → Lac operon is on • Lactose = inducer, turns operon on

An **operon** is a functioning unit of genomic DNA containing a cluster of genes under the control of a single promoter. The genes often have similar functions, or help with different parts of a larger process.

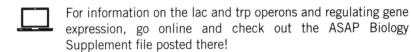 For information on the lac and trp operons and regulating gene expression, go online and check out the ASAP Biology Supplement file posted there!

Applying Molecular Biology

Scientists began experimenting with recombinant DNA and applied molecular biology techniques in the 1960s and 1970s, and have come a long way since then. These experiments allow scientists to manipulate and analyze DNA, RNA, and proteins. Genetic engineering techniques can manipulate the heritable information of DNA, and even produce new organisms.

Definitions

Recombinant DNA

- Made by combining DNA from two different species
- Makes unique DNA that is not found in nature
- New genetic combinations can be valuable to science, medicine, agriculture, and industry

Genomic DNA (gDNA)

- DNA in the chromosomes of an organism
- The same genomic DNA is found in every cell of a certain organism

Remember!

Only certain genes are expressed in each cell, which allows different cells to have different structures and functions.

Genetic Engineering

- Modifying the genome of an organism using biotechnology
- Often done by transferring genes between cells of the same or different species
- Produces improved or new organisms

Cloning ●

- Producing populations of genetically identical DNA fragments, cells, or organisms

Reagents ●

Plasmids

- Double-stranded and circular DNA
- Usually smaller than a chromosome
- Extrachromosomal (non-essential)
- Self-replicating
- Can be found in many prokaryotes and some eukaryotes
- Used in molecular biology labs to transfer and replicate DNA

One important characteristic of the plasmids used in molecular biology labs is that they give a transformed cell **selective advantage**. For example, most plasmids contain an antibiotic resistance gene such as Amp^R, which allows bacteria with the plasmid to survive when ampicillin is added to the growth culture. Bacterial cells without the plasmid will die in the presence of ampicillin. This is a clever way scientists can find out which bacteria have taken up a plasmid.

Bacterium

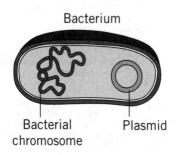

Bacterial chromosome   Plasmid

Restriction Enzymes

- Enzymes that cut double-stranded DNA at specific sites (called recognition sites)
- Can generate compatible "sticky ends" on the DNA

 Remember!

Restriction enzymes are endonucleases; they cut DNA in the middle of a strand. In contrast, exonucleases cut DNA at the end of a nucleotide chain.

DNA Ligase

- An enzyme that connects (or ligates) two pieces of DNA together
- Catalyzes the formation of a phosphodiester bond

 Remember!

Restriction enzymes cleave phosphodiester bonds, and DNA ligase forms them.

For an example of how restriction enzymes and ligase work, go online and check out the ASAP Biology Supplement file posted there!

Many restriction enzymes are isolated from species of bacteria and they are often named for the organism they come from. For example, *BamHI* is a restriction enzyme derived from *Bacillus amyloliquefaciens,* and *EcoRI* was isolated from *Escherichia coli.*

The restriction sites of restriction enzymes (the sites they recognize and cut) are usually palindromes. A nucleotide sequence is said to be palindromic if it is the same as its reverse complement (the complementary sequence read backwards). For example, the DNA sequence 5' – ACCTAGGT – 3' is palindromic because its nucleotide-by-nucleotide complement is 3' – TGGATCCA – 5'; reversing the order of the nucleotides in the complement gives the original sequence. *EcoRI* recognizes the restriction site 5' - GAATTC - 3'.

Complementary DNA (cDNA)

- Double-stranded DNA synthesized from a single-stranded RNA
- Made by the enzyme reverse transcriptase
- Often used to clone eukaryotic genes in prokaryotes
- cDNA of a gene has no introns, so is shorter than the genomic DNA sequence of the same gene

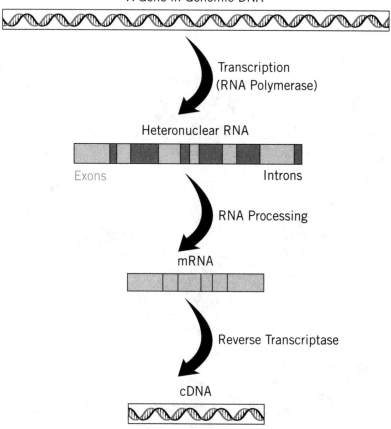

A Gene in Genomic DNA

Transcription
(RNA Polymerase)

Heteronuclear RNA

Exons   Introns

RNA Processing

mRNA

Reverse Transcriptase

cDNA

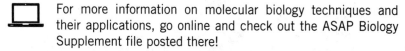

For more information on molecular biology techniques and their applications, go online and check out the ASAP Biology Supplement file posted there!

 cDNA and reverse transcriptase represent an exception to the central dogma of molecular biology. Normally information flows from DNA to RNA, but here, it flows in the opposite direction.

Viruses !

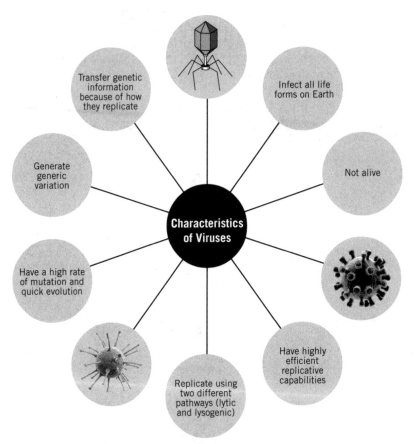

- All viruses have an outer protein coat, called the capsid.
- The genome is inside, and it can be:
  - Made of DNA or RNA
  - Single or double-stranded
  - Circular or linear

Ask Yourself...

Why does ^{32}P radioactively label the viral genome? Why does ^{35}S radioactively label the viral capsid?

Are viruses alive?	• No!
Are viruses cells?	• No!
They are:	• Obligate intracellular parasites • Infective particles
Because they:	• Can't perform any of the chemical reactions characteristic of life (e.g. synthesis of ATP and macromolecules) • Cannot survive without a host • Cannot reproduce without a host • Have no activity outside host cells

Remember!

- Obligate intracellular parasites are only able (obligated) to reproduce within (intra) cells.

- In the final analysis, a virus is nothing more than a package of nucleic acid that says: "Pick me up and reproduce me!"

Viruses can replicate using one of two pathways: lytic or lysogenic.

Lytic Cycle ❗

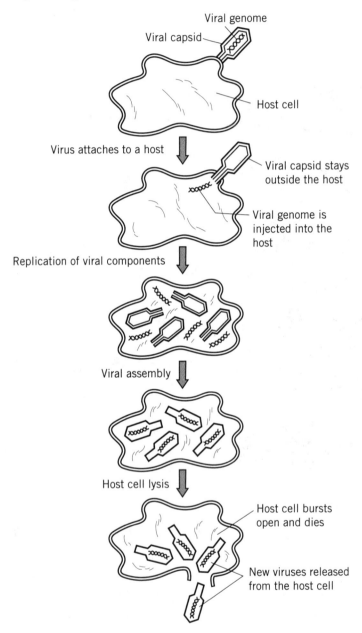

Viral genome

Viral capsid

Host cell

Virus attaches to a host

Viral capsid stays outside the host

Viral genome is injected into the host

Replication of viral components

Viral assembly

Host cell lysis

Host cell bursts open and dies

New viruses released from the host cell

Lysogenic Cycle 🛑

- Viral genome stays "undercover" in the host genome
- This means the viral genome is secretly replicated every time the host genome is replicated
- This is a good system for the virus, as long as the host is healthy and dividing
- If the host cell becomes unfavourable, the viral genome cuts out of the host genome and enters the lytic cycle instead

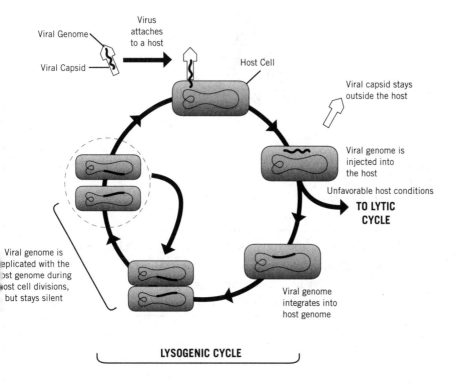

Viral Genome

Viral Capsid

Virus attaches to a host

Host Cell

Viral capsid stays outside the host

Viral genome is injected into the host

Unfavorable host conditions

TO LYTIC CYCLE

Viral genome is replicated with the host genome during host cell divisions, but stays silent

Viral genome integrates into host genome

LYSOGENIC CYCLE

Transduction

- When the viral genome cuts out of the host genome, it can take some host genome with it
- Virus carries this host DNA with virus DNA
- Virus can pass old host DNA onto the next host
- In other words, viruses can transfer genetic information between hosts—a process called transduction

? Ask Yourself...

Is there an upper limit to how much host DNA a virus can carry? Why?

Viruses are very specific in the host they use. Animal viruses must overcome additional challenges not present in bacterial cells. Some examples are immune pathways, organelle structures, and different cell machinery.

Retroviruses !

- RNA viruses that replicate using the lysogenic cycle
- Use the enzyme reverse transcriptase to convert their RNA genomes into DNA, so it can be inserted into a host genome
- Example: human immunodeficiency virus (HIV)

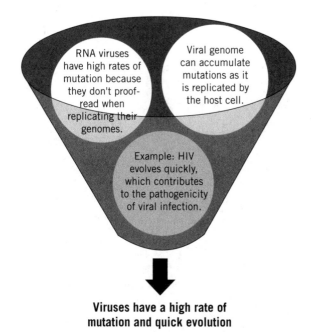

RNA viruses have high rates of mutation because they don't proof-read when replicating their genomes.

Viral genome can accumulate mutations as it is replicated by the host cell.

Example: HIV evolves quickly, which contributes to the pathogenicity of viral infection.

Viruses have a high rate of mutation and quick evolution

CHAPTER 5

Heredity and Genetics

Modern genetics was founded by a monk studying inheritance patterns in peas. Hundreds of years later, we know what chromosomes look like and how they and the genes and alleles they encode cause phenotypes, variations, and disorders.

Mendelian Genetics ❗

Gregor Mendel was the father of genetics; he described how traits in pea plants were inherited long before we knew about DNA and chromosomes. Although Mendelian genetics generally only involves the simplest patterns of inheritance, it forms the foundation for understanding more complicated genetics.

Gregor Mendel

Scientist and monk from 19th century	Founder of the modern science of genetics	Performed pea plant experiments	Established many rules of heredity, now referred to as the laws of Mendelian inheritance

 Mendel began his experiments on mice, but switched to gardening after a request from his Bishop. Many phenotypes in mice have complex inheritance patterns, so without the switch, Mendel may never have figured out his rules of inheritance.

Genetics

- Study of genes, genetic variation, and heredity in living organisms
- Explains how traits are passed from parents to children

Trait

- Expressed characteristic or quality
- Genetically determined

Gene

- Unit of hereditary
- Passed from parent to offspring
- Segment of a chromosome
- Includes: promoter, RNA-coding sequence, terminator

 Ask Yourself...

Do all genes code for a protein?

Genotype	• Genetic makeup of an organism • Which alleles the organism has
Phenotype	• Physical appearance of an organism • What the organism looks like
Dominant	• Denoted with a capital letter • Trait that is expressed in a heterozygote • Cannot be masked by another allele
Recessive	• Denoted with a lowercase letter • Trait that is not expressed by a heterozygote • Can be masked by another allele
Homozygous	• When an organism has two identical alleles for a given gene or trait
Heterozygous	• When an organism has two different alleles for a given gene or trait

Genotype vs. Phenotype

PHENOTYPE			
GENOTYPE	*BB* Homozygous Dominant	*Bb* Heterozygous	*bb* Homozygous Recessive

For some examples of Mendelian traits in humans, go online and check out the ASAP Biology Supplement file posted there!

Haploid vs. Diploid ❗

Haploid (*n*)

- When a cell or oganism has one set of chromosomes
- The haploid number for humans is 23 (*n* = 23)

Diploid (2*n*)

- When a cell or organism has two sets of chromosomes (one inherited from its mother and one from its father)
- The diploid number for humans is 46 (2*n* = 46)

 Ask Yourself...

1. Are all organisms diploid?
2. Are there any options other than haploid and diploid?

Chromosome vs. Chromatid 💬

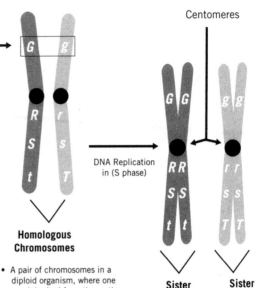

Alleles at a genetic locus →

Locus
- Position or location of a gene on a chromosome

Alleles
- Different forms of the same gene
- Can be dominant or recessive

<u>Examples:</u>
G and *g* alleles of the G gene
R and *r* alleles of the R gene
S and *s* alleles of the S gene
T and *t* alleles of the T gene

Centomeres

DNA Replication in (S phase)

Homologous Chromosomes

- A pair of chromosomes in a diploid organism, where one was inherited from the mother and the other from the father
- Same size
- Same structure
- Same genes
- Same or different alleles
- Similar but not identical sequences

Sister Chromatids **Sister Chromatids**

- Result from a duplicated chromosome
- Same size
- Same structure
- Connected at a centromere
- Same genes
- Same alleles
- Identical sequences

Genetic Crosses and Generations

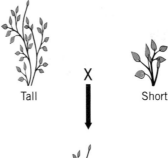

P generation

- Parental generation
- The first generation in an experiment

Tall X Short

F₁ generation

- First filial generation
- Offspring of the P generation

All tall

Self-fertilization

F₂ generation

- Second filial generation
- Grandchildren of the P generation
- Offspring of the F₁ generation

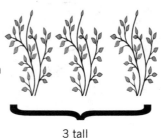

3 tall   1 short

Monohybrid Cross ❗

A monohybrid cross is a mating between two individuals at one genetic locus of interest. A Punnett square can be used to predict the outcome of a particular cross or breeding experiment.

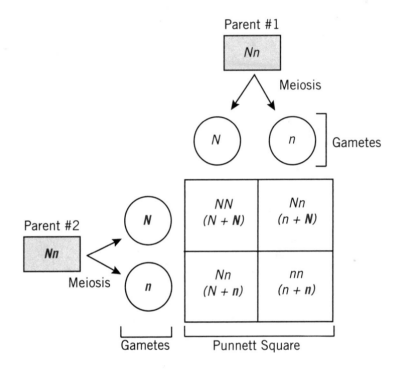

Totals:

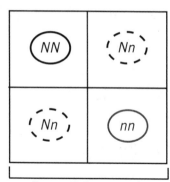

1 ⟮NN⟯

2 ⟮Nn⟯

1 ⟮nn⟯

Punnett Square of
Offspring **Genotypes**

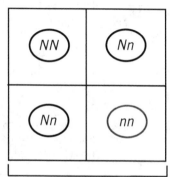

3 ⟮N–⟯

1 ⟮nn⟯

Punnett Square of
Offspring **Phenotypes**

A monohybrid cross between two heterozygote individuals results in a
genotypic ratio of 1:2:1 and a phenotypic ratio of 3:1.

Dihybrid Cross ❗

A dihybrid cross is a mating between two individuals at two genetic loci of interest.

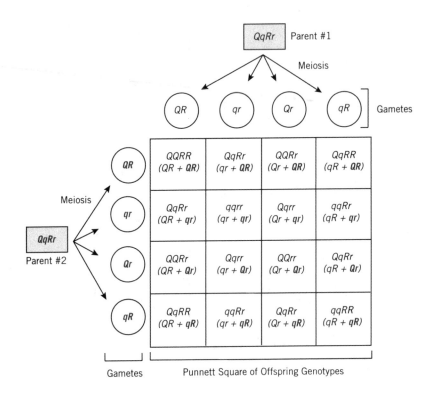

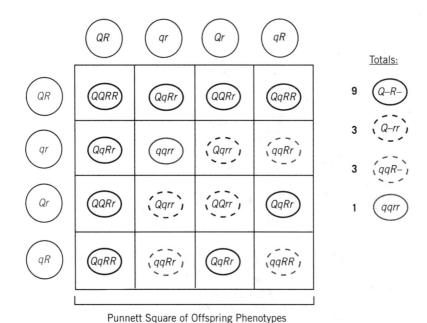

Punnett Square of Offspring Phenotypes

A dihybrid cross between two heterozygote individuals results in a phenotypic ratio of 9:3:3:1. Use the rules of probability to calculate genotypic ratios.

Testcross ❗

To determine the genotype of an individual, Mendel developed a procedure called a testcross. The testcross is performed by taking an individual of unknown genotype, and crossing them with a homozygous recessive individual.

Let's start with two peas of unknown genotype (*A* and *B*) for roundness (*R*) or wrinkledness (*r*). They are both round in phenotype, so their genotype must be *RR* or *Rr*. The testcross allows you to determine which of these genotypes is true, based on offspring data:

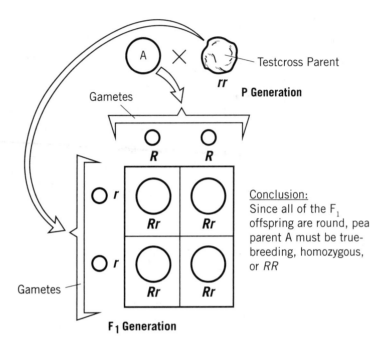

Testcross Parent

P Generation

Gametes

Conclusion:
Since all of the F₁ offspring are round, pea parent A must be true-breeding, homozygous, or *RR*

Gametes

F₁ Generation

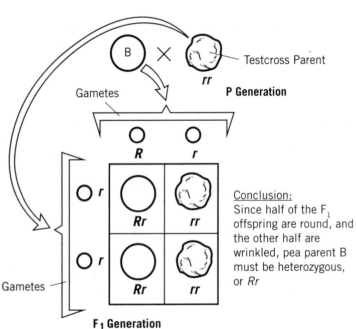

Testcross Parent

P Generation

Gametes

Conclusion:
Since half of the F₁ offspring are round, and the other half are wrinkled, pea parent B must be heterozygous, or *Rr*

Gametes

F₁ Generation

Probabilities in Genetics Problems

For complex crosses, a Punnett square is cumbersome and time consuming. Instead, learn the four monohybrid crosses below, and use the rules of probabilities.

Monohybrid Crosses		
Cross	**Genotyping Ratio**	**Phenotyping Ratio**
$AA \times AA$	100% $AA = 1$	100% $A = 1$
$AA \times aa$	100% $Aa = 1$	100% $A = 1$
$AA \times Aa$	50% $AA = \dfrac{1}{2}$ 50% $Aa = \dfrac{1}{2}$	100% $A = 1$
$Aa \times Aa$	25% $AA = \dfrac{1}{4}$ 50% $Aa = \dfrac{1}{2}$ 25% $aa = \dfrac{1}{4}$	75% $A = \dfrac{3}{4}$ 25% $a = \dfrac{1}{4}$
$Aa \times aa$	50% $Aa = \dfrac{1}{2}$ 50% $aa = \dfrac{1}{2}$	50% $A = \dfrac{1}{2}$ 50% $a = \dfrac{1}{2}$
$aa \times aa$	100% $aa = 1$	100% $a = 1$

 Remember!

The probability of a male offspring is $\dfrac{1}{2}$, and the probability of a female offspring is $\dfrac{1}{2}$.

Product Rule ❗

- Also called the "and rule"
- Probability of independent events occurring together is the product of their separate probabilities

Sum Rule 💬

- Also called the "either-or-rule"
- Probability of either of two events occurring is the sum of their individual probabilities

Mendelian Genetics Example

A woman heterozygous for brown eyes (blue are recessive) and brown hair (blond is recessive) mates with a man with the same genotype. What is the probability they will have a son with brown hair and brown eyes who is able to have blue-eyed children himself?

Suppose:

- H = brown hair and h = blond hair
- E = brown eyes and e = blue eyes

P Generation:

- $EeHh \times EeHh$

Want to find the probability of a male with H–Ee

- $P(\text{male}) = \dfrac{1}{2}$

- $P(H-) = \dfrac{3}{4}$

- $P(Ee) = \dfrac{1}{2}$

Overall, $P(\text{male with } H\text{–}Ee) = \dfrac{1}{2} \times \dfrac{3}{4} \times \dfrac{1}{2} = \dfrac{3}{16}$

 **Remember!**

Probability can be expressed as a fraction, percentage, or a decimal.

❗	Mendel's Laws: 3 Principles of Genetics	
Law of Dominance	One trait can mask the effects of another trait	
Law of Segregation	Alleles can segregate and recombine	
Law of Independent Assortment	Traits can segregate and recombine independently of other traits	

Creating New Phenotypes

Changes in genotypes can result in changes in phenotype, but environmental factors also influence many traits, both directly and indirectly. See the chart on the following page.

Height and Weight in Humans
- Controlled by both genetic and environmental factors
- **Examples:** nutrition, disease, physical activity, medication use

Soil pH Affects Flower Color
- Soil pH affects nutrient accessibility, which can change flower color
- In acidic soil (pH<5.5), aluminum is available to the roots of hydrangeas, resulting in blue flowers
- When no aluminum is available, flowers are white

Seasonal Fur Color in Arctic Animals
- Arctic hare is brown or gray in the summer, when their habitat is full of greens and browns
- Arctic hare is white in winter because everything is covered in snow
- Color change is linked to photoperiod (or how much light is received during the day)

Sex Determination in Reptiles
- Temperatures experienced during embryonic/larval development determines the sex of offspring
- In lizards, male embryos develop at intermediate temperatures and female embryos are generated at both extremes

Herbivores Affect Density of Plant Hairs
- Plant hairs interfere with the feeding of some herbivores and insects
- Damage from insects or herbivores causes an increase in hair density on plants

Gene Expression when Bacteria are Grown in Presence of Lactose
- Presence of lactose allows expression of the lac operon
- Lac operon codes for enzymes that digest lactose
- These genes are not expressed if lactose is absent from growth media

UV Increases Melanin Production in Animals
- Melanin contributes to the pigmentation of hair, skin and eyes
- UV rays penetrate skin and affect melanin production in melanocytes
- Contributes to camouflage, thermoregulation and UV protection

Hormones Control Mating in Yeast
- Yeasts release diffusible sex hormones called pheromones
- Bind to other yeast cells of the same species
- Cause a signaling cascade that results in changes in cell structure, to facilitate mating

💬 An organism's adaptation to the local environment reflects a flexible response of its genome. For example, gene expression and protein function can be affected by temperature.

Siamese Cats Have Partial Albinism (Pale Body) and Darker Extremities
- Results from a mutation in tyrosinase, an enzyme involved in melanin production
- Mutated enzyme fails to work at normal body temperatures, leading to albinism (no color)
- Mutated enzyme becomes active in cooler areas of the skin, so extremities accumulate dark pigment

Climate Change Affects the Timing of Flowering
- Elevated temperatures affect the physiology of flowering plants
- **Examples:** altered production of flowers, nectar, and pollen
- Can have a ripple effect on other parts of the ecosystem
- **Examples:** insects that pollinate plants or the birds that eat them

❗ Multiple copies of alleles or genes (gene duplication) may produce new phenotypes.

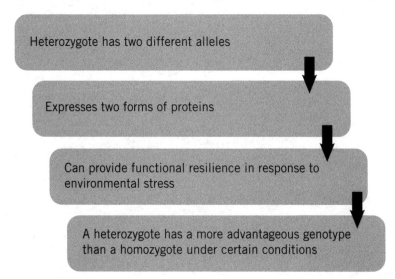

Heterozygote has two different alleles

Expresses two forms of proteins

Can provide functional resilience in response to environmental stress

A heterozygote has a more advantageous genotype than a homozygote under certain conditions

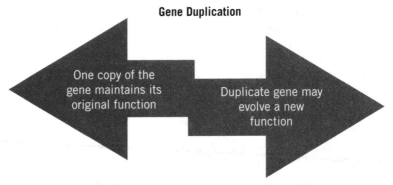

Gene Duplication

One copy of the gene maintains its original function

Duplicate gene may evolve a new function

Effect: Multiple copies of alleles or genes may produce new phenotypes

Beyond Mendelian Genetics

Not all patterns of inheritance obey the principles of Mendelian genetics. In fact, many traits we observe are due to a combined expression of genes and alleles. Also, some traits are determined by genes on sex chromosomes and others result from non-nuclear inheritance.

Many traits are the product of multiple genes or alleles, cooperating in complex ways.

Codominance

- Alleles are neither dominant nor recessive

- Heterozygote expresses both alleles (not blended)

Incomplete Dominance

- Alleles are neither dominant nor recessive

- Phenotype of a hetero-zygote is a blended mix of both alleles

Multiple Alleles

- Many genes have more than two alleles possible

- Remember: Diploid organisms have one or two of these

Polygenism

- Phenotype is determined by many genes

- These traits tend to display a range of pheno-types in a continuous distribution

Linkage ❗

- Two genes on the same chromosome and less than 50 map units apart do not segregate indepen-dently

- Inherited together, as a unit

- The closer two genes are, the less recombination (or cross-ing over) occurs be-tween them

Incomplete Dominance: Example

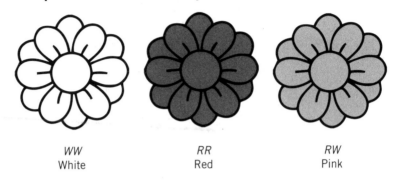

WW
White

RR
Red

RW
Pink

 Ask Yourself...

If the *R* allele is dominant to the *W* allele, what would be the phenotype of a *RW* individual?

Codominance and Multiple Alleles: Example 🗨

- There are 3 alleles for human ABO blood type
- I^A and I^B are codominant
- i is recessive (and non-coding)

Genotype	ii	$I^A i$ $I^A I^A$	$I^B i$ $I^B I^B$	$I^A I^B$
RBC				
Phenotype (Blood Type)	O	A	B	AB

 Ask Yourself...

1. Which blood type is the universal donor?
2. Which blood type is the universal acceptor?
3. How can blood type be used to answer questions of paternity?

ASAP Biology

Polygenism: Example 💬

- Skin color in humans is controlled by at least 3 genes
- Each has a "dark" dominant allele and a "light" recessive allele
- More dark alleles = darker skin: see (#) in chart below
- More light alleles = lighter skin

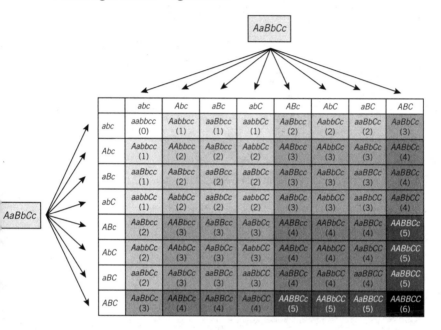

	abc	Abc	aBc	abC	ABc	AbC	aBC	ABC
abc	aabbcc (0)	Aabbcc (1)	aaBbcc (1)	aabbCc (1)	AaBbcc (2)	AabbCc (2)	aaBbCc (2)	AaBbCc (3)
Abc	Aabbcc (1)	AAbbcc (2)	AaBbcc (2)	AabbCc (2)	AABbcc (3)	AAbbCc (3)	AaBbCc (3)	AABbCc (4)
aBc	aaBbcc (1)	AaBbcc (2)	aaBBcc (2)	aaBbCc (2)	AaBBcc (3)	AaBbCc (3)	aaBBCc (3)	AaBBCc (4)
abC	aabbCc (1)	AabbCc (2)	aaBbCc (2)	aabbCC (2)	AaBbCc (3)	AabbCC (3)	aaBbCC (3)	AaBbCC (4)
ABc	AaBbcc (2)	AABbcc (3)	AaBBcc (3)	AaBbCc (3)	AABBcc (4)	AABbCc (4)	AaBBCc (4)	AABBCc (5)
AbC	AabbCc (2)	AAbbCc (3)	AaBbCc (3)	AabbCC (3)	AABbCc (4)	AAbbCC (4)	AaBbCC (4)	AABbCC (5)
aBC	aaBbCc (2)	AaBbCc (3)	aaBBCc (3)	aaBbCC (3)	AaBBCc (4)	AaBbCC (4)	aaBBCC (4)	AaBBCC (5)
ABC	AaBbCc (3)	AABbCc (4)	AaBBCc (4)	AaBbCC (4)	AABBCc (5)	AABbCC (5)	AaBBCC (5)	AABBCC (6)

AaBbCc (top)
AaBbCc (left)

Ask Yourself...

If two heterozygotes for skin color mate, what is the probability of having a child with no dark alleles? 1 dark allele? 2 dark alleles? 3 dark alleles? 4 dark alleles? 5 dark alleles? All 6 dark alleles?

Linkage: Example ❗

Independent Assortment:
- Genes are on the same chromosome but more than 50 map units apart
- Recombination or crossing over happens normally
- Equal probabilities of allele combinations in gametes

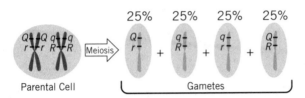

Linked Genes:
- Genes are on the same chromosome but less than 50 map unit apart
- Recombination or crossing over happens less often
- Unequal probabilities of allele combinations in gametes

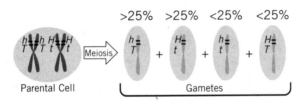

Remember!

The probability two genes will segregate as a unit is a function of the distance between them.

Ask Yourself...

When does crossing over occur in meiosis? Does it occur in mitosis?

For another example of linked genes, go online and check out the ASAP Biology Supplement file posted there!

Inheritance of Sex Chromosomes

Some traits are determined by genes on sex chromosomes. Others are determined by genes on autosomes.

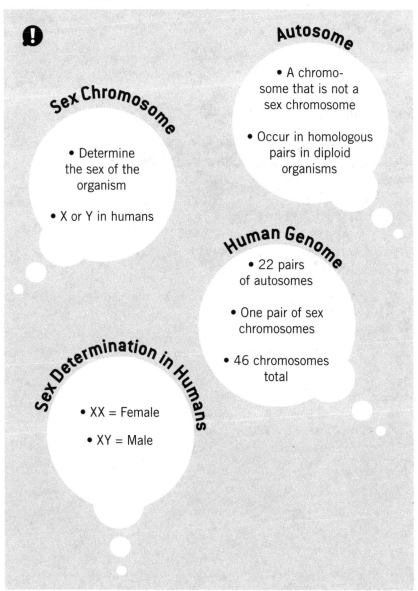

Sex Chromosome
- Determine the sex of the organism
- X or Y in humans

Autosome
- A chromosome that is not a sex chromosome
- Occur in homologous pairs in diploid organisms

Human Genome
- 22 pairs of autosomes
- One pair of sex chromosomes
- 46 chromosomes total

Sex Determination in Humans
- XX = Female
- XY = Male

Remember!

Germline mutations are heritable and passed to offspring via the gametes.

Remember!

Somatic mutations are acquired and not inherited from a parent. They can affect the individual but are not passed to offspring.

Sex Chromosomes in Humans 💬

	X Chromosome	Y Chromosome
Size	Large	Small
Genes	Many	Few
# in Females	2	0
# in Males	1	1

Remember!

Human X and Y chromosomes are not homologous: they are different sizes and encode different genes.

Ask Yourself...

Can crossing over or recombination occur between two X chromosomes? Between an X and a Y chromosome?

Flies (such as the fruit fly, *Drosophila melanogaster*) also use the XY sex-determination system.

Birds use a ZW sex-determination system instead of XY. Male birds are ZZ and females are ZW. The Z chromosome is larger and has more genes (like the X chromosome in the XY system).

ASAP Biology

Sex Chromosome Inheritance in Humans

- Mothers pass an X chromosome to all offspring
- Fathers pass an X chromosome to all daughters
- Fathers pass a Y chromosome to all sons
- Daughters inherit an X chromosome from each parent
- Sons inherit an X chromosome from their mother and a Y chromosome from their father

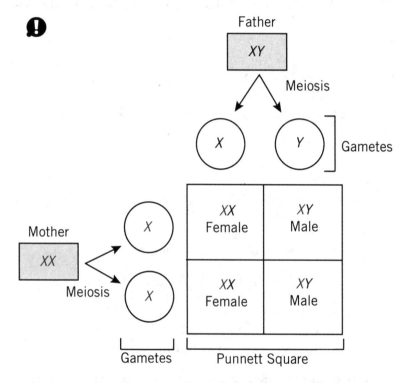

X-Linked Inheritance

X-linked dominant traits are always expressed by both males and females. They cannot be masked by a normal allele.

X-linked recessive traits are always expressed by males (because they have only one X chromosome). They can be masked in females by a normal/unaffected allele. In other words, women can carry X-linked recessive traits but not express them.

💬	X-Linked Dominant Traits	X-Linked Recessive Traits
Affected Genotypes	X^AX^A X^AX^a X^AY	X^aX^a X^aY
Unaffected or Normal Genotypes	X^aX^a X^aY	X^AX^A X^AX^a X^AY
Prevalence	Less common	More common

X-Inactivation

- Early in mammalian embryonic development, one X chromosome is inactivated in females
- Occurs at random
- Ensures males and females have the same dose of genes on the X chromosome
- Inactive X chromosome in a female somatic cell = Barr Body
- Human female = one Barr body per somatic cell
- Human male = no Barr bodies

Examples of X-Linked Dominant Traits

Trait	Gene	Phenotype
Rett Syndrome	*MECP2*	Postnatal neurological disorder of the grey matter of the brain
Vitamin D Resistant Rickets	*PHEX*	Bone deformity including short stature and bow leggedness
Fragile X Syndrome	*FMR1*	Developmental problems including learning disabilities and cognitive impairment

Examples of X-Linked Recessive Traits

Trait	Gene	Phenotype
Hemophilia	*Clotting Factor VIII or IX*	Impaired blood clotting, bruising easily
Red/Green Color Blindness	*Opsin 1*	Difficulty discerning red and green hues

 For more examples of X-linked recessive traits, go online and check out the ASAP Biology Supplement file posted there!

 Remember!

Many of these conditions can be caused by random (*de novo*) mutations instead of being caused by mutations inherited from either parent.

Autosomal Traits Affected by Sex

Sex-Limited Traits	Sex-Influenced Traits
Expressed in one sex but not the other	Expressed differently in the two sexes

Women have a gene for heavy beard growth; don't express it but can pass it onto sons	Men have a gene for mammary milk production; don't express it but can pass it onto daughters	Breast cancer occurs less frequently in males and more frequently in females	Male pattern baldness is a dominant trait in males and a recessive trait in females

Autosomal Inheritance in Humans

- Mothers pass one of each autosome to all offspring
- Fathers pass one of each autosome to all offspring
- Dominant alleles, traits, or mutations will be expressed and cannot be masked
- Recessive alleles, traits, or mutations will be expressed only if an individual carries two copies; they can be masked by dominant alleles.

 Remember!

Autosomal dominant traits cannot skip generations. Autosomal recessive traits can.

 Remember!

"Skipping a generation" means an individual in the P generation expresses the trait, their F_1 offspring do not (but can carry the allele), and an individual in the F_2 generation expresses the trait again.

Examples of Autosomal Traits

Trait	Gene	Genotype	Phenotype
Huntington's Disease	Autosomal Dominant	Varying length of trinucleotide repeats in the HTT gene	Gradual death of brain cells, causing physical limitations and eventual dementia
Sickle-Cell Anemia	Autosomal Recessive	Point mutation in the hemoglobin gene	Blood disorder due to altered shape of red blood cells
Tay-Sachs Disease	Autosomal Recessive	Mutation of the HEXA gene	Gradual death of nerve cells, causing deterioration of mental and physical abilities

 Ask Yourself...

An individual expressing an autosomal recessive trait must carry how many mutant alleles?

Non-Nuclear Inheritance

Not all traits are inherited via genes and chromosomes in the nucleus.

Endosymbiosis Theory

- Evolutionary theory of the origin of eukaryotic cells
- Mitochondria and chloroplasts used to be free-living prokaryotes
- Around 1.5 billion years ago, they were taken inside a cell

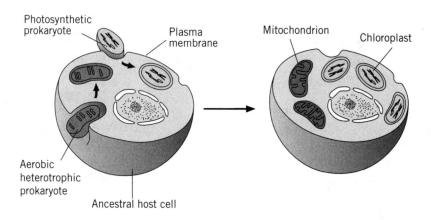

Support

Mitochondria, chloroplasts, and bacterial cells all:

- Divide/form through binary fission
- Have similar membrane structure
- Contain one circular double-stranded DNA molecule

Mitochondrial DNA is sometimes short formed as mtDNA and chloroplast DNA is called cpDNA.

The human genome contains 3.2 billion base pairs. Mitochondrial DNA is double-stranded, circular and 16,569 base pairs in size.

Application to Inheritance 💬

- During cell division, chloroplasts, and mitochondria assort randomly into daughter cells and gametes
- Mitochondrial traits and DNA:
  - Transmitted by the egg, not sperm
  - Maternally inherited: affected mother passes the trait to all offspring
  - Not Mendelian

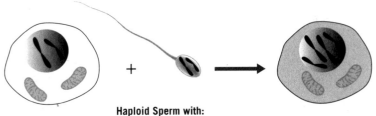

Haploid Sperm with:
- 1 set of nuclear chromosomes

Haploid Ovum with:
- 1 set of nuclear chromosomes
- Mitochondrial DNA

Diploid Zygote:
- 2 sets of nuclear chromosomes
- Mitochondrial DNA from ovum

❓ Ask Yourself...

1. It is commonly said that humans inherit 50% of their DNA from their mother and 50% from their father. Given what you know about the inheritance of mitochondrial DNA, is this true?

2. Leber's hereditary optic neuropathy is a mitochondrial trait and leads to vision loss in young adults. Will a woman with this mutation always express it, or can it be hidden? What about a man with this mutation? If the woman breeds, what proportion of her offspring will inherit the trait? What about the man?

3. What kind of processes do you think the genes on mitochondrial and chloroplast DNA code for?

CHAPTER 6

Evolutionary Biology

All organisms around today arose from other organisms. Evolution is viewed at the species level, but changes happen to individuals.

Natural Selection 😮

Charles Darwin provided an evidence-based mechanism for evolution, called natural selection.

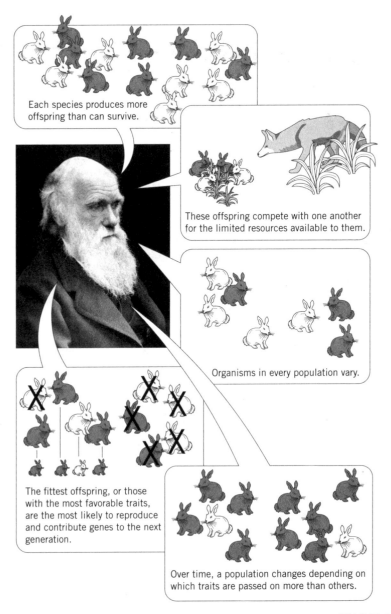

Each species produces more offspring than can survive.

These offspring compete with one another for the limited resources available to them.

Organisms in every population vary.

The fittest offspring, or those with the most favorable traits, are the most likely to reproduce and contribute genes to the next generation.

Over time, a population changes depending on which traits are passed on more than others.

Evolutionary Fitness ❗

Evolutionary Fitness is the ability to survive and create offspring in a particular environment.

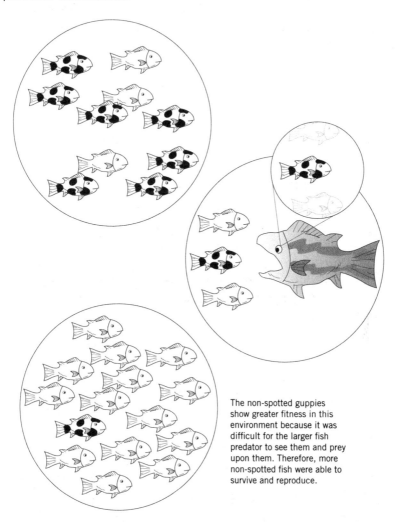

The non-spotted guppies show greater fitness in this environment because it was difficult for the larger fish predator to see them and prey upon them. Therefore, more non-spotted fish were able to survive and reproduce.

 Ask Yourself...

1. Why do each of Darwin's criteria have to be met for natural selection to occur?
2. What is the difference between evolutionary fitness and common every day usage?

Genetic Variation is Essential for Natural Selection ❗

For species to evolve there must be genetic variability that causes differences in phenotype. This comes from random mutation and new allele combinations, recombination, and gene transfer.

If everyone is the same, then nobody has an advantage to reproduce better and nobody is naturally selected. Variation is a requirement for natural selection.

Mutation

Errors in DNA replication and repair can lead to mutation. These mutations are constantly occurring at random and can be beneficial, harmful,
or neutral, depending on the gene they are changing and how much it
changes the phenotype. Tiny variations in phenotype are always appear-
ing. If you need to review mutations, go back to Chapter 4.

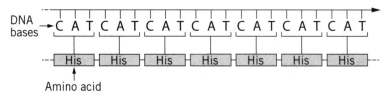

Original DNA code for an amino acid sequence.

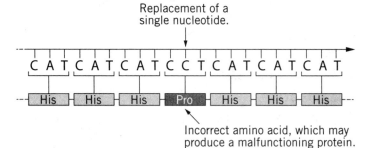

Ask Yourself...

Why doesn't evolution act on genotypes?

Recombination 💬

In sexual reproduction, genetic variation arises from the mixing of parental DNA. Plus, during gamete formation, homologous chromosomes recombine and exchange gene segments, which creates new combinations of alleles. This will be discussed more in the next chapter.

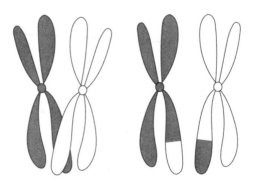

Gene Transfer 💬

In asexual reproduction, genetic information can be passed by bacterial conjugation or viral transduction to contribute to genetic variation.

Natural Selection Acts on Phenotypes ❗

Natural selection acts on phenotypes, not genotypes. A genotype describes a genetic recipe; it is a segment of DNA locked away in the cell, and is not an observable trait. The phenotype (observable characteristic) made from that recipe is being naturally selected. This is because the phenotype can give an advantage or disadvantage to the organism, in terms of its survival and/or reproduction.

Environmental Selective Pressure ❗

Genetic variation provides the differences in phenotype, but there must be an environmental factor that makes some traits advantageous and others disadvantageous. This is called a selective pressure.

Advantageous Phenotypes are Selected by Environmental Conditions

Recombination and mutation lead to genetic variation, which causes phenotype variations, such as the color varieties in moths. However, which color is advantageous (has the most evolutionary fitness) depends on the environment.

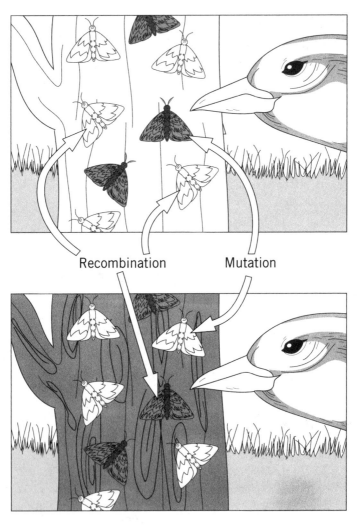

Recombination   Mutation

Ask Yourself...

How might hand sanitizer use today shape the bacteria of the future?

Selection Results in Evolutionary Change

Antibiotics kill most bacteria but if some bacteria have a phenotypic variation that allows them to survive, those will be the only ones that pass on their genes to the next generation.

Thus, the bacterial population will evolve to only include bacteria that have resistance to the antibiotics, the selecting agent.

Streptococcus pneumonia, the bacteria causing pneumonia, was previously treated with penicillin. Now, almost no strains of this bacteria can be treated with penicillin. This is strong evidence for the fact that evolution is still occurring today.

Before selection

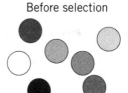

After selection

Final population

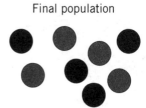

Resistance level

Evidence for Evolution ❗

Scientific evidence has been shown to indicate that all living things have evolved from one common ancestor and evolution continues to occur. What evidence you ask? THIS EVIDENCE!

Biogeographical Evidence ❗

Biogeography is the study of species distribution together with geological evidence of what the Earth used to look like. Isolated species that used to geographically coexist are closely related. Species in long geographically-isolated areas are quite unique.

Galapagos tortoises and Chaco Tortoise (*Geochelone chilensis*), a small tortoise found on the nearby South American mainland, have been closely linked through mitochondrial DNA.

Galapagos Islands

Equador

Molecular Evidence

DNA—Extremely similar DNA sequences have been found in all domains of life demonstrating that all life has evolved from common ancestors.

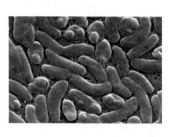  

Molecular Evidence

Protein—Sequencing amino acids in collagen, the main protein in connective tissue, collected from T. Rex remains was compared to chicken collagen and showed many similarities.

Collagen

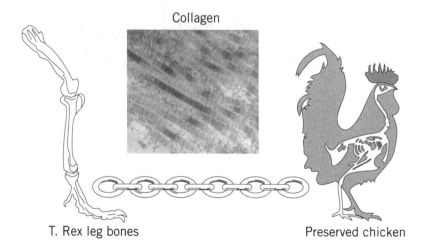

T. Rex leg bones   Preserved chicken

Phylogenetic Trees ❗

Data gathered from genetic and amino acid sequencing, as well as morphological similarities in fossil evidence, have allowed organisms to be linked through phylogenetic trees. These diagrams show the inferred evolutionary relationship between species. On phylogenetic trees, the branches represent evolutionary time. These trees are created through computer analysis of these data.

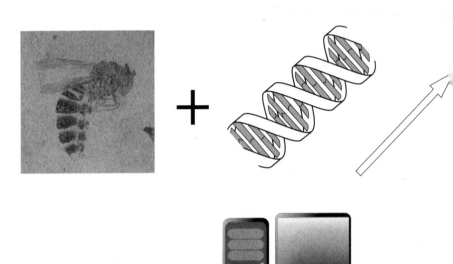

Phylogenetic Tree of Life

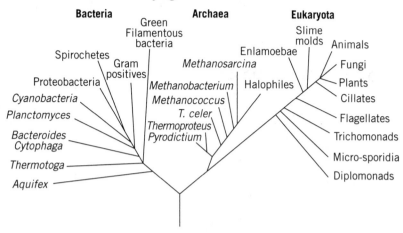

Each time one line splits from another, it means a divergence event has occurred. In other words, one ancestral group was split into two descendant groups.

 Ask Yourself...

Which evidence provides the strongest case for evolution?

Comparative Anatomy 🔴

Comparative anatomy is the study of anatomical similarities in different animals.

Homologous structures 🔴

Homologous structures are anatomical features that are evolutionarily similar, but presently serve a different purpose.

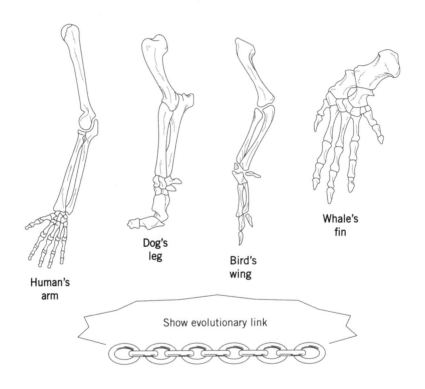

Human's arm

Dog's leg

Bird's wing

Whale's fin

Show evolutionary link

Analogous Structures ❗

Analogous structures are anatomical features that work similarly, but do not show an evolutionary connection.

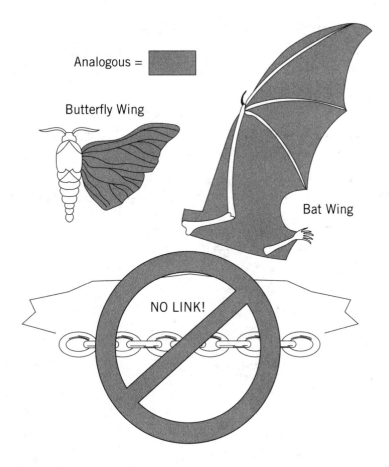

Analogous =

Butterfly Wing

Bat Wing

NO LINK!

❓ Ask Yourself...

If you are unsure if a trait is analogous or homologous, what tests might you do to figure this out?

Beginnings of Life and Speciation Events !

The Origins of Life on Earth !

Approximately 4 billion years ago, life appeared on Earth. It is believed that the early Earth likely contained water (H_2O), methane (CH_4), ammonia (NH_3), hydrogen (H_2), and possibly other molecules.

In 1952, the Miller-Urey experiment showed that amino acids could be produced by adding electrical sparks to an environment with a mixture of molecules thought to be present in the early atmosphere.

Before this point, it is possible that life (as most define it) began with an RNA genome since RNA can have enzymatic properties and self-replicate.

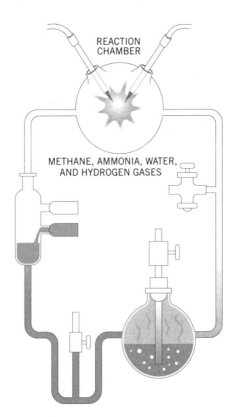

REACTION
CHAMBER

METHANE, AMMONIA, WATER,
AND HYDROGEN GASES

| Early Earth | Organic molecules | RNA world with self-replicating RNA | DNA genome and protein enzymes |

Evolution in Motion 💬

Evolution occurs either due to natural selection or genetic drift.

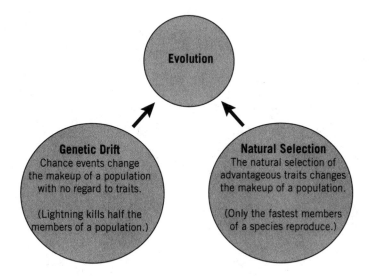

Evolution

Genetic Drift
Chance events change the makeup of a population with no regard to traits.

(Lightning kills half the members of a population.)

Natural Selection
The natural selection of advantageous traits changes the makeup of a population.

(Only the fastest members of a species reproduce.)

Adaptations 💬

Adaptive traits are functional characteristics that were naturally selected due to the advantage they provide. Many species have a specific niche habitat in which they survive, and the development of special characteristics that have allowed them to live there is adaptation. Species are so different because of the many paths by which they have adapted.

Speciation ❗

If members of a species become reproductively isolated, each population will continue undergoing natural selection. Depending on the genetic variation and the selective pressures in each population, different traits can emerge and become naturally selected. If the two species become so different that they can no longer reproduce, they are now considered different species.

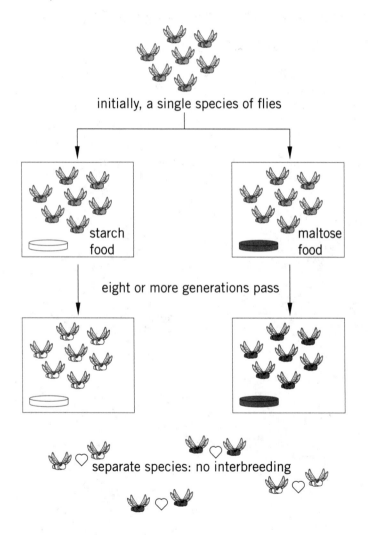

Population Genetics ❗

The foundation of evolution is that the genetic makeup of a species changes over time. Population genetics is the study of the genotype frequencies within a population.

Genetic Diversity Increases the Survival Rate of a Population 💬

A species with more genetic diversity is more likely to have individuals that can survive a given selective pressure. The likelihood of extinction is decreased with increased genetic variation, since a threat to one individual is less likely to be equally threatening to all individuals.

Threat to all.

Just the spotted beetles are attacked.

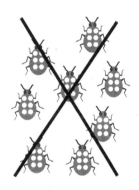

Environmental Instability Increases Evolution Rates 💬

Extreme selective pressures hasten natural selection as the unfit individuals will be removed from the population more quickly. In bacterial resistance, a single antibiotic can kill 99.999% of bacteria in one treatment, leaving only the resistant bacteria to survive and reproduce.

Of course, certain selective pressures may be too extreme to overcome. Earth has seen several mass extinction events, leaving only the hardiest life forms to evolve into future generations.

Hardy-Weinberg Equilibrium ❗

If a population could be "closed," it is said to be in Hardy-Weinberg Equilibrium and the frequency of alleles will remain unchanged as the individuals within the population mate and reproduce with each other. However, the equilibrium will be broken unless each of the following conditions for a "closed" population is met.

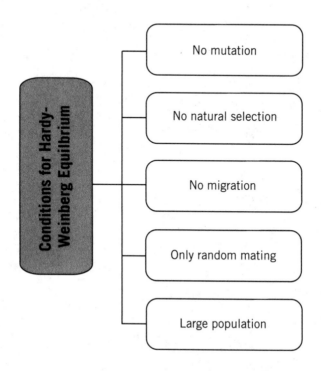

Hardy-Weinberg Equations ❗

If the gene of interest only comes in two alleles that show classical dominance, then the following equations can be used to identify the allele frequencies and the genotype frequencies for the population.

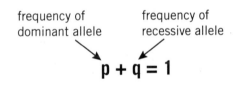

frequency of dominant allele

frequency of recessive allele

$$p + q = 1$$

$$p^2 + 2pq + q^2 = 1$$

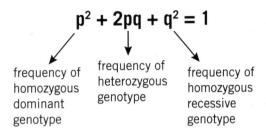

frequency of homozygous dominant genotype

frequency of heterozygous genotype

frequency of homozygous recessive genotype

Ask Yourself...

A population of 1200 aardvarks have a gene coding for snout length that comes in two alleles (long and short). 200 aardvarks have the short recessive phenotype. How many aardvarks are heterozygotes?

CHAPTER 7

Cell Reproduction

Living things are able to produce offspring. The basis of this is cell reproduction, which in humans is accomplished by mitosis and meiosis. Controlled cell division is also important in growth and wound healing, while uncontrolled cell division is one of the hallmarks of cancer.

The Cell Cycle ❗

The life cycle of a cell is called the cell cycle, and includes four phases.

G₁ Phase

- Gap phase 1
  - Growth
- Metabolism
- Protein synthesis
- Getting ready for S phase

S Phase

- Synthesis of the genome
- DNA replication and repair

G₂ Phase

- Gap phase 2
  - Growth
- Metabolism
- Protein synthesis
- Getting ready for M phase

M Phase

- Mitosis
- Includes prophase, metaphase, anaphase, telophase, cytokinesis

The Gs in G₁ and G₂ stand for "Gap" because under a low-power microscope, they look like a gap in cell activity (as opposed to S and M phases). In actual fact, lots is going on in these phases!

Remember!

G_1, S, and G_2 phases are together referred to as interphase.

Ask Yourself...

1. In what phase of the cell cycle would new organelles be made?

2. Would helicase and primase be made primarily in G_1 or G_2? Why?

S Phase and DNA Replication 🔴

You learned about DNA replication in Chapter 4. It duplicates chromosomes so they contain two identical chromatids instead of one (also see Chapter 5). Sister chromatids are held together at the centromere.

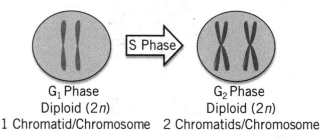

G_1 Phase
Diploid (2*n*)
1 Chromatid/Chromosome

S Phase

G_2 Phase
Diploid (2*n*)
2 Chromatids/Chromosome

Remember!

Counting centromeres is a good way to count chromosomes. This can help you determine if a cell is diploid (has a double set of chromosomes) or haploid (has a single set of chromosomes).

In a typical rapidly proliferating human cell, the cell cycle lasts about 24 hours; G_1 phase is about 11 hours, S phase is about 8 hours, G_2 is about 4 hours, and M takes about 1 hour. In contrast, a typical bacteria can divide every 20 minutes!

Cell Cycle Control 💬

The cell cycle is closely controlled and regulated by checkpoint pathways and Cyclin/CDK proteins.

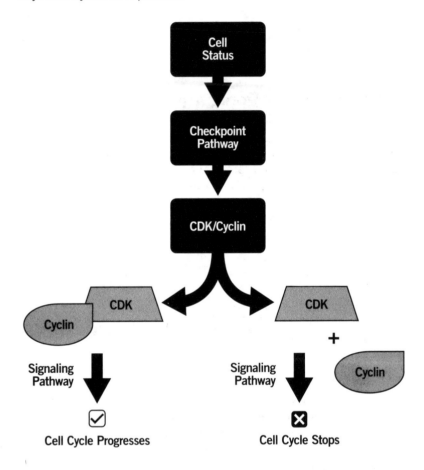

Cell Cycle Checkpoints 💬

- Internal controls
- Monitor the cell to make sure it is ready to progress through the cell cycle
  - If so, they promote the cell cycle
  - If not, they stop cell cycle progression
- Control associations between Cyclin and CDK proteins

Remember!

Cyclins are regulatory proteins that control CDKs.

Ask Yourself...

CDKs associating with Cyclins is an example of what level of protein structure?

Some Cells Don't Cycle

Cells don't have to always actively cycle. Cells can exit the cell cycle to die via apoptosis, or to become a more specialized (or differentiated) non-cycling cell.

Exiting the Cell Cycle

Apoptosis

- Programmed cell death
- Limits damage to neighbor cells
- Normal part of development and health in multicellular organisms

Senescence

- Cell is alive and metabolically active
- Cell does not actively divide
- Irreversible; the cell cannot re-enter the cell cycle
- Examples: red blood cells, cells on the surface of your skin

Quiescence or G_0

- Cell is alive and metabolically active
- Cell does not actively divide
- Reversible; the cell can re-enter the cell cycle
- Examples: nerves and heart muscle cells

The Cell Cycle ❗

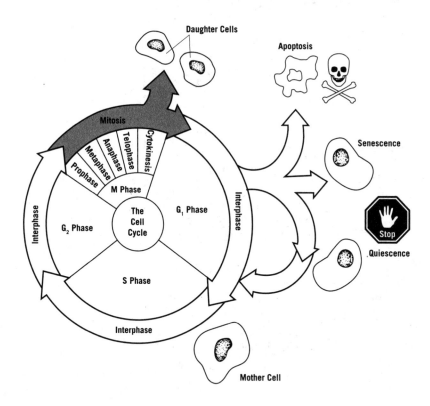

Cancer 💬

Cancer occurs when normal cells start behaving and growing very abnormally, and spread to other parts of the body.

How a normal cell transforms into a cancer cell:

- Override cell cycle check-points
- Grow in an unregulated way
- Avoid cell death (apoptosis)
- Accumulate DNA damage
- Spread to other parts of the body

Cancer treatments target these changes, since they are what makes a cancer cell different from a normal cell.

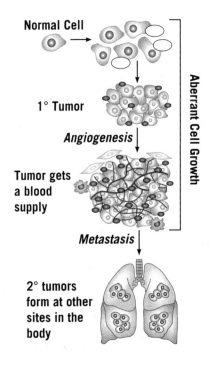

Normal Cell

1° Tumor

Angiogenesis

Tumor gets a blood supply

Aberrant Cell Growth

Metastasis

2° tumors form at other sites in the body

 Remember!

Angiogenesis is the formation of new blood vessels, and helps tumors obtain a blood supply.

 Remember!

Metastasis is the spread of cancer cells to new areas of the body, usually via the lymph system or bloodstream.

 Cancer means "crab," as in the zodiac sign. The name comes from the observation that malignant tumors grow into surrounding tissue, embedding themselves like clawed crabs. Cancer is not one disease; it is a collection of hundreds of diseases.

ASAP Biology

Ask Yourself...

A blood supply gives tumor cells a way to spread to other parts of the body. Why is obtaining a blood supply important for tumors to keep growing at the original site?

😶 Cancer cells commonly have mutations in two types of genes: oncogenes and tumor suppressor genes.

	Oncogene	Tumor Suppressor Gene
Function	Promote or induce cancer when mutated	Prevent the conversion of normal cells into cancer cells
In the cell	Promotes cell growth and survival	Inhibits the cell cycle Promotes apoptosis
In cancer	☑ ON: Activated by mutation Over-expressed Amplified	☒ OFF: Inactivated by mutation Gene expression turned off Deleted

Remember!

You learned about mutations in Chapter 4. Review that content again now if you need a refresher.

Over-expression means too much protein is made from a gene in the genome. Amplification means the gene is duplicated so there are too many copies in the genome, but each makes the normal amount of protein. Both lead to too much of a certain protein in the cell.

"Onco-" is a prefix denoting cancer

 **Remember!**

Overall, the mutations in a cancer cell turn on growth signals and turn off death signals.

 **Ask Yourself...**

In breast cancer, the *TP53* gene often has an inactivating mutation, while the *PIK3CA* gene often has an activating point mutation. Is *TP53* an oncogene or a tumor suppressor gene? What about *PIK3CA*?

Mitosis

Many eukaryotic cells use mitosis to replicate cells. This reproduction is asexual and increases the number of cells in a multicellular eukaryotic organism.

Ask Yourself...

What cell process decreases the number of cells in an organism?

Machinery of Mitosis

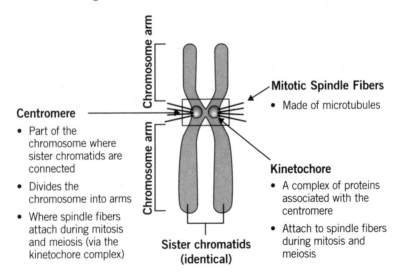

Chromosome arm

Mitotic Spindle Fibers
- Made of microtubules

Centromere
- Part of the chromosome where sister chromatids are connected
- Divides the chromosome into arms
- Where spindle fibers attach during mitosis and meiosis (via the kinetochore complex)

Chromosome arm

Kinetochore
- A complex of proteins associated with the centromere
- Attach to spindle fibers during mitosis and meiosis

Sister chromatids (identical)

Remember!

Telomeres are at the end of a chromosome; don't mix these up with centromeres!

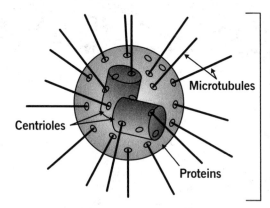

Centrosome

- An organelle made of two centrioles and many proteins
- Replicated during the S phase
- Anchors mitoic spindle fibers (microtubules)

 Remember!

Be careful with centromeres, centrosomes, and centrioles! These are all different structures even though the words are so (annoyingly?) similar!

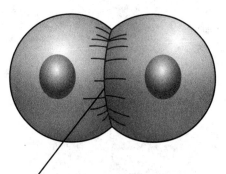

- Cytokinesis is accomplished by cleavage furrow formation and contraception
- The cleavage furrow is made of a ring of microfilaments encircling the cell
- Microfilaments are made of actin

⚠️ Mitosis passes a complete genome from the parent cell to daughter cells. It is a continuous process with observable structural features:

Phase	What It Looks Like	What is Happening?	Cell Ploidy	Chromatids per Chromosome
Interphase (G$_1$, S and G$_2$)	Nucleus containing Chromatin Nuclear Membrane — Nucleolus Cytoplasm Centrioles Centrosome	• DNA replication • Centrosome replication • Preparing for mitosis	2n	1 → 2
Early Prophase	Replicated Chromosome (with 2 sister chromatids) Asters — Cytoplasm Centrioles — Nuclear Membrane Centrosome Centrioles Asters	• Chromatin condenses to form chromosomes • Nucleolus disappears • Mitotic spindle fibers appear and form asters	2n	2
Late Prophase	Cytoplasm — Spindle Fibers Chromosome with 2 Sister Chromatids Nuclear Membrane Centrosome	• Nuclear envelope degrades • Centrosomes migrate to opposite ends of the cell	2n	2
Metaphase	Cytoplasm — Centrioles in Centrosome Spindle Fibers — Kinetochore Metaphase Plate Chromosomes (each has 2 sister chromatids)	• Chromosomes line up on metaphase plate • Mitotic spindle fibers attach to kinetochores	2n	2

hase	What It Looks Like	What is Happening?	Cell Ploidy	Chromatids per Chromosome
Anaphase	 Kinetochore — Cytoplasm; Chromosomes (each has 1 sister chromatid); Spindle Fibers; Centrioles in Centrosome	• Chromosomes separate • Sister chromatids pulled to opposite ends of the cell • Microtubules in spindle fibers shorten • Cell elongates	$2n$ ↓ $4n$	$2 \rightarrow 1$
Telophase and Cytokinesis	Centrioles; Centrosome; Nucleolus; Nuclear Membrane; Cleavage Furrow; Spindle Fibers; Chromosomes (each has 1 sister chromatid); Cytoplasm	• Nuclear envelope and nucleolus reform • Chromosomes relax to chromatin • Cytokinesis forms two identical diploid daughter cells	$4n$ ↓ $2n$	1

Remember!

In most human cells, cytokinesis starts near the end of anaphase and is completed during telophase.

Remember!

Cytokinesis occurs differently in plant cells. Instead of a cleavage furrow, a partition called a cell plate forms down the middle of the dividing cell.

Remember!

Most of the cells in your body are called somatic cells; these cells are diploid and divide via mitosis.

Mitosis: The Dance of the Chromosomes

- DNA replication doubles the number of chromatids per chromosome (1 → 2)
- Chromosome separation in anaphase doubles the number of chromosomes ($2n$ → $4n$) but halves the number of chromatids per chromosome (2 → 1)
- Cytokinesis halves the number of chromosomes per cell ($4n$ → $2n$)

Remember!

In mitosis, the parental cell has the same genome as both daughter cells.

Remember!

One way to remember the phases of mitosis (**P**rophase, **M**etaphase, **A**naphase, **T**elophase) is the mnemonic **P**eople **M**eet **A**nd **T**alk.

Ask Yourself...

When in mitosis does the cell have the highest number of centromeres?

Here is a cool trick to help you remember what happens when in mitosis:

Phase	Hint	Explanation
Prophase	Prepare	The cell gets ready to divide
Metaphase	Middle	Chromosomes line up in the middle of the cell
Anaphase	Apart	Chromosomes are pulled apart and sister chromatids separate
Telophase	Two	The parental cell splits into two daughter cells

Meiosis

In Chapter 5 you learned that humans inherit half of their nuclear chromosomes from their mother and half from their father: haploid gametes fuse in fertilization to form a diploid zygote. Meiosis generates these haploid gametes.

Mendel's Law of Independent Assortment requires that genes or traits on the same chromosome segregate and recombine independently. You learned about this in Chapter 5 and will learn more about how it happens here.

 Ask Yourself...

What are Mendel's other laws?

💬 Crossing over occurs in meiosis (prophase I) but not in mitosis

  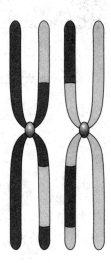

Synapsis: Homologous Chromosomes in a tetrad   **Crossing Over**   **Recombined chromosomes**

Remember!

The prefix "tetra" means 4.

Remember!

A tetrad contains four chromatids.

Meiosis involves two rounds of cell division: meiosis I and meiosis II.
Before meiosis begins, the genome is replicated (just like in mitosis).
This generates two identical sister chromatids per chromosome.

Meiosis I ❶

Many details of meiosis I are similar to mitosis. Here is a summary of the differences:

Phase	What It Looks Like	Differences from Mitosis	Cell Ploidy	Chromatids per Chromosome
Early Prophase I	Nucleolus, Cytoplasm, Chromatid, Nucleus, Centrosomes	• None	2n	2
Late Prophase I	Centrosome, Spindle Fibers, Tetrad: 2 homologous chromosomes with 2 chromatids each	• Synapsis and crossing over occur	2n	2
Metaphase I	Sister Chromatids, Homologous Chromosomes, Metaphase Plate, Sister Chromatids	• Chromosomes line up on either side of the metaphase plate	2n	2

hase	What It Looks Like	Differences from Mitosis	Cell Ploidy	Chromatids per Chromosome
Anaphase I	**2 Sister Chromatids in one Chromosome** **2 Sister Chromatids in one Chromosome**	• Homologous chromosomes move to opposite ends of the cell • Sister chromatids are NOT pulled apart	$2n$	2
Telophase I and Cytokinesis	**Cleavage Furrow** **2 Sister Chromatids in one Chromosome** **2 Sister Chromatids in one Chromosome**	• Cytokinesis forms two haploid daughter cells • Each chromosome still contains 2 sister chromatids	$2n$ ↓ $1n$	2

Ask Yourself...

Does DNA replication occur between meiosis I and meiosis II?

Meiosis II !

The second half of meiosis is even more similar to mitosis in terms of how it happens. However, meiosis II starts with a haploid cell and generates haploid gametes (or sex cells). Egg and sperm cells are examples of gametes.

Phase	What It Looks Like	Differences from Mitosis	Cell Ploidy	Chromatids per Chromosome
Prophase II	Centrioles in Cytoplasm Centrosome, 2 Sister Chromatids in one Chromosome, 2 Sister Chromatids in one Chromosome	• None	$1n$	2
Metaphase II	2 Sister Chromatids in one Chromosome, 2 Sister Chromatids in one Chromosome, Metaphase Plate	• None	$1n$	2
Anaphase II	Chromosome 1, Chromosome 3, Chromosome 2, Chromosome 4	• None	$1n$ ↓ $2n$	$2 \rightarrow 1$
Telophase II and Cytokinesis	Chromosome 1, Chromosome 3, Cleavage Furrow, Cleavage Furrow, Chromosome 2, Chromosome 4	• Cytokinesis forms four haploid gametes	$2n$ ↓ $1n$	1

Ⓠ **Remember!**

Gamete formation is called gametogenesis.

❓ **Ask Yourself...**

What is the name of gametogenesis in human males? Females? Where do these processes occur?

Products of Meiosis 💬

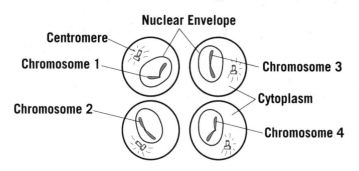

![Centromere, Chromosome 1, Chromosome 2, Nuclear Envelope, Chromosome 3, Cytoplasm, Chromosome 4]

Ask Yourself...

In mitosis, cells have a tetraploid (4n) moment in anaphase. Does this occur in meiosis?

Ask Yourself...

When in meiosis do cells become haploid?

Remember!

Meiosis and fertilization contribute to genetic diversity in sexually reproducing organisms.

Ask Yourself...

When in meiosis does the cell have the highest number of centromeres?

Nondisjunction

Sometimes chromosome separation in meiotic anaphase doesn't happen properly: either homologous chromosomes don't move apart, or sister chromatids fail to separate. This is termed nondisjunction, and can lead to gametes with too many or not enough chromosomes. When these gametes undergo fertilization, monosomic and trisomic zygotes can form.

Remember!

Diploid cells have two copies of each chromosome and are thus called "disomic." Cells that are missing a certain chromosome are called monosomic, and cells with an extra copy of a certain chromosome are called trisomic.

Aneuploidy is the presence of an abnormal number of chromosomes in a cell (for example a human cell having 45 or 47 chromosomes instead of the usual 46). It is a major category of chromosome mutations and is the cause of many birth defects.

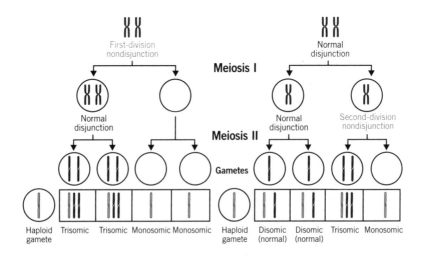

Certain human genetic disorders can be attributed to nondisjunction.

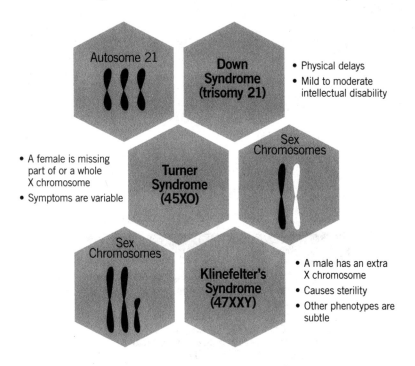

- Physical delays
- Mild to moderate intellectual disability

- A female is missing part of or a whole X chromosome
- Symptoms are variable

- A male has an extra X chromosome
- Causes sterility
- Other phenotypes are subtle

Two other autosomes can undergo trisomy in humans, but babies born with these mutations usually have severe birth defects, disabilities, and abnormalities in many parts of the body. Trisomy 13 is also called Patau Syndrome, and trisomy 18 is also known as Edwards Syndrome. Most other autosome trisomies are lethal.

Aneuploidy of the sex chromosomes is pretty common and compatible with life.

Many ethical, social, and medical issues surround human genetic disorders:

Reproduction Issues

- Individuals who are carriers for genetic diseases are at increased risk to have a child with a genetic disease
- Prenatal diagnosis can identify some genetic diseases before birth
- Parents can then prepare to have a child with special needs, or have the option of terminating the pregnancy
- Individuals who carry genetic diseases can use an egg or sperm donor instead

Civic Issues

- Genetic discrimination occurs when people are treated unfairly because of actual or perceived differences in their genetic information that may cause or increase the risk of developing a disorder or disease
- For example, a health insurer might refuse to give coverage to an individual who has a genetic difference that raises her odds of getting cancer
- Many G7 countries protect genetic test information to help eliminate genetic discrimination and safeguard genetic privacy

Ask Yourself...

Go back to the mitosis summary chart on pages 164 and 165. What is the haploid number for this cell? What would 2n equal? What about 4n?

For a summary of ploidy, go online and check out the ASAP Biology Supplement file posted there!

CHAPTER 8

Animal Structure and Function

Multicellular organisms are complex. Interaction and coordination between specialized systems is key to maintaining homeostasis and the overall functioning of an organism.

Specialized Body Systems Interact ❗

Body systems contain specialized tissues and organs that perform specific functions, but they cannot function without coordination with other body systems. The entire body works together to maintain homeostasis.

For instance, the respiratory system exchanges gases with the environment. The circulatory system then transports these gases to and from various tissues. These two systems rely on one another to provide each cell with oxygen and to eliminate CO_2 waste.

Respiratory System

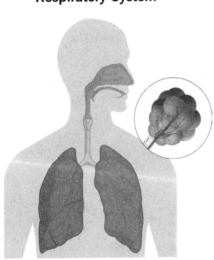

Exchanges Gases

Circulatory System

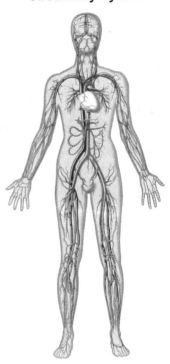

Carries gases and other substances to other body tissues

Main Body Systems 💬

For the AP exam you only need to memorize details of the immune system, the nervous system, and the endocrine system, but here is a quick guide to the functions of the other major systems. You might see passages about them or get a chance to use your knowledge in a free response question.

Body System	Function
Circulatory	Carries substances throughout the body
Respiratory	Exchanges gases
Digestive	Converts food into energy and nutrients
Excretory	Removes wastes from the body
Reproductive	Produces gametes and nourishes offspring
Muscular	Moves the body and substances within the body
Skeletal	Provides support and protection to the body; red blood cell production
Immune	Defends the body
Nervous	Transmits signals throughout the body
Endocrine	Produces hormones to regulate the body

 Ask Yourself...

How do different systems need to interact to maintain homeostasis?

The Immune System 🛑

The body responds to natural or artificial agents that disrupt homeostasis. If these invaders are biological, such as bacteria or viruses, they are called pathogens.

Non-specific Responses 🛑

Invertebrates and vertebrates have first-line immune responses that are non-specific. These include physical barriers, inflammation, complement proteins, and scavenger cells.

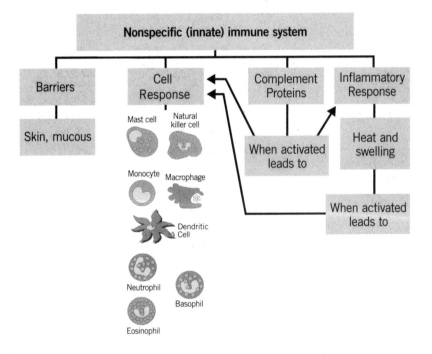

Specific Immune Reponses

Mammals use specific immune responses which adapt as they are exposed to different pathogens. The two types of specific response pathways are humoral and cell-mediated.

Humoral 🔋

This is led by immune cells circulating out in the body's fluids (AKA the body's humors). These cells travel through the blood and lymphatic system, but they especially like to hang out in lymph nodes, which swell when an active battle is taking place there.

B-cells are the stars of the humoral response. B-cells have Y-shaped receptors on their membranes called **antibodies** that bind to **antigens**, which are bits of foreign invader.

All of the arms on an antibody bind the same antigen, but each antibody has a slightly different shape, called the variable region, so each antibody binds to a different antigen.

With the help of Helper T-cells (discussed next), the B-cells develop into either:

- Plasma B-cells that release antibodies out into the body to seek out and attach to invaders, marking them for destruction.
- Memory B-cells that wait around like a militia in case the pathogen ever enters the body again. A second exposure to an antigen results in a more rapid and enhanced immune response because the memory B-cells are waiting and ready for that specific pathogen.

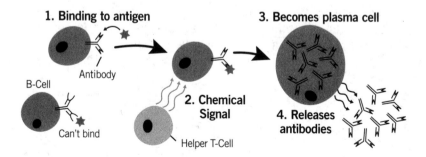

1. Binding to antigen
Antibody
B-Cell
Can't bind
2. Chemical Signal
Helper T-Cell
3. Becomes plasma cell
4. Releases antibodies

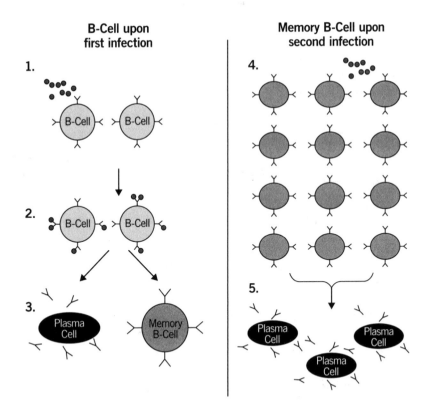

B-Cell upon first infection

1.

2.

3.

Memory B-Cell upon second infection

4.

5.

Cell-mediated ⚠

These immune responses are dependent on direct cell-to-cell interactions. The stars of this are T-cells working together with cells that have special cell-surface proteins called major histocompatibility complexes (MHC).

- MHC is a group of proteins that sits atop cells and displays things that are found within the cell. All regular cells have MHC I on them and certain special immune cells have MHC II.
- T-cells work together with the MHC to check out what they are displaying.
  - Cytotoxic T-cells: They inspect what MHC I has found INSIDE our cells and if it is bad they mount an attack. Examples: Cancer, viruses, intracellular bacteria.
  - Helper T-cells: They inspect what MHC II on B-cells (and other immune cells) has found and if it is bad then they activate the B-cell.

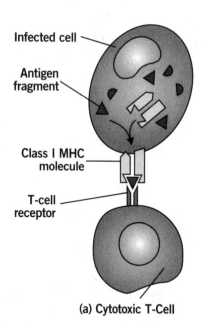

Infected cell

Antigen fragment

A fragment of foreign protein (antigen) inside the cell associates with an MHC molecule and is transported to the cell surface.

Class I MHC molecule

T-cell receptor

The combination of MHC molecule and antigen is recognized by a T-cell, alerting it to the infection.

(a) Cytotoxic T-Cell

AIDS

HIV is a particularly dangerous virus because it attacks Helper T-Cells. Without Helper T-cells, B-cells will not be activated. If they are not activated they cannot create plasma cells and memory B-cells. Therefore, infections that would easily be fought off by the specific immune system lead to potentially deadly consequences.

HIV Progression

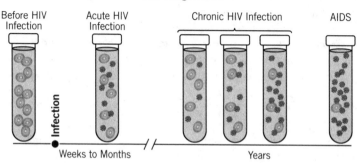

Before HIV Infection

Acute HIV Infection

Chronic HIV Infection

AIDS

Infection

Weeks to Months

Years

= Helper T-cells   = HIV

The Nervous System ❗

The nervous system collects information, makes decisions, and orders responses using electrical signals sent throughout the body.

Neurons ❗

Neurons are the cells of the nervous system.

Parts of the Neuron ❗

- Cell body—the central area where most of the organelles of the neuron are housed.
- Axon—the long portion of the neuron that sends the signal to other cells (nerves are always the axon portion of the cell) (Remember: Axon away from cell body)
- Dendrites—receive signals from other cells (Remember: Dendrites deliver to cell body)

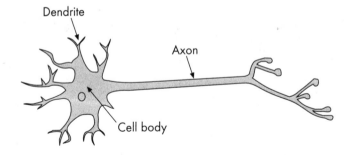

Myelin sheath: Provides insulation for an axon so that signals can be sent quickly over long distances.

- Schwann cells wrap themselves around the axon to make the myelin sheath in the central nervous system
- Oligodendrocytes secrete the myelin sheaths in the peripheral nervous system.

Parts of the Nervous System 💀

Central Nervous system (CNS) ❗

The brain and the spinal cord comprise the central nervous system. The CNS receives signals from the peripheral nervous system, processes the signals, and then sends a signal back to the peripheral nervous system to cause an effect.

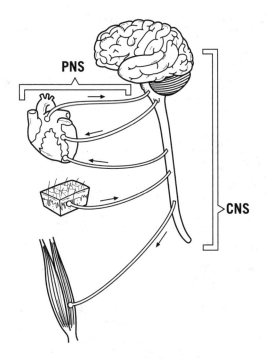

 The sciatic nerve is the longest nerve in the human body extending from the base of the spinal cord to the foot.

Peripheral Nervous System (PNS) 💬

The peripheral nervous system is any part of the nervous system that is not part of part of the brain and spinal cord. The nerves that go to and from your body parts are in this system.

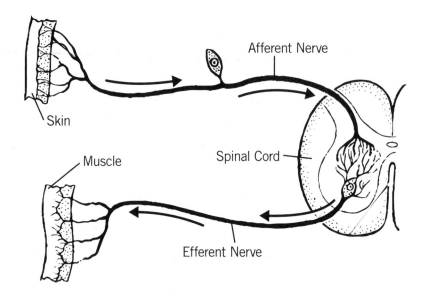

Types of Neurons 💬

- Sensory neurons are in the PNS and they take a stimulus from an organ and send that signal to the CNS (e.g. skin receptors sending a temperature reading). Also called afferent neurons.
- Interneurons are in the CNS and are a relay between neurons.
- Motor neurons are in the PNS and they send a signal from the CNS to an effector (e.g. muscle cell) to cause an effect (e.g. shivering). Also called efferent neurons.

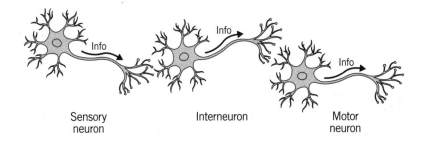

Sensory
neuron

Interneuron

Motor
neuron

How Neurons Communicate 🔊

Neurons communicate through electrical signaling. The transmitted signal is called an action potential.

Resting Potential 🔊

All cells have a negative charge on the inside compared to the outside. This is due to differences in ion concentrations inside and outside the cell set up by the sodium-potassium pump and leaky K^+ channels. This causes more positive ions to leave the cell than to enter it. The resting membrane potential of the cell is $-70mV$, which means the cell is polarized (more negative inside).

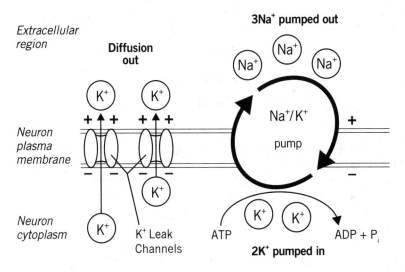

 **Ask Yourself...**

What would happen to the resting membrane potential if the same number of K^+ and Na^+ were pumped?

Action Potential ❗

In short, action potentials are a wave of positivity inside the cell caused by opening and closing of different ion channels.

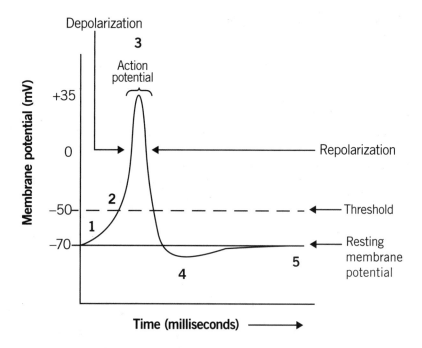

1. A stimulus reaches the cell and causes it to become slightly more positive
2. When it becomes so positive that it reaches −50mV (called threshold) voltage-gated Na^+ channels open, Na^+ rushes IN, and the inside of the cell becomes very positive very quickly. This is called depolarization because the cell is less polarized (i.e. it isn't super negative anymore).

3. When the cell becomes so positive that it reaches +35mV, the Na⁺ channels close and the cell stops getting more positive. At the same time, voltage-gated K⁺ channels open, K⁺ rushes OUT of the cell, and it starts becoming negative again.
4. When the cell is so negative that it reaches –90mV, the K⁺ channels close and the cell stops getting more negative. During the period when the cell is even more negative than when it started the cell is hyperpolarized because it is even more polarized than usual.
5. Eventually, the resting membrane potential is restored by the sodium potassium pump.

 Ask Yourself...

What would happen if a cell's Na⁺ channels were blocked?

All-Or-Nothing

A stimulus brings the cell closer to threshold (stimulatory) or further from threshold (inhibitory), but it plays no role in the actual action potential. Once threshold is reached, the action potential will proceed the same way each time since the maximum positivity and negativity are determined by the ion channels and not by the original stimulus. For this reason, action potentials are said to be all-or-none. They either happen or they don't, but the action potential itself does not change.

The Refractory Period

In neurons, an action potential cannot fire again right away for two reasons:

1. Na⁺ channels need a rest period before they can open again.
2. K⁺ channels don't close until the cell is hyperpolarized, so it takes more positive stimulus to bring the cell to threshold.

This delay is called the refractory period, and ensures that an action potential completes a full cycle and returns to the resting state before firing again.

Synapses ❗

The zone where two neurons connect is called a synapse. The axon of one neuron connects to a dendrite of another neuron (or sometimes a final axon will connect to a muscle or another organ). The axon will release tiny neurotransmitter molecules into the synapse that diffuse over to the next cell.

If they are stimulatory they will cause the cell to become more positive and push it toward threshold. If they are inhibitory they will cause the cell to become hyperpolarized. Acetylcholine, norepinephrine, and GABA are some common neurotransmitters.

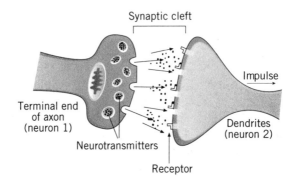

The Endocrine System ❗

The endocrine system is a way to pass chemical messages through the bloodstream to distant tissues.

Hormones ❗

- Chemical messages in the endocrine system are called hormones.
- They are released from glands and carried through the blood until they reach their target tissues/effector cells

---

 The toxin in a black widow spider causes a huge release of the neurotransmitter, acetylcholine, which causes systemic pain, rigidity, and sweating.

❗ Here's a summary of some of the hormones and their effects on the body.

Organ	Hormones	Effect
Anterior Pituitary	FSH	Stimulates activity in ovaries and testes
	LH	Stimulates activity in ovary (release of ovum) and production of testosterone
	ACTH	Stimulates the adrenal cortex
	Growth Hormone	Stimulates bone and muscle growth
	TSH	Stimulates the thyroid to secrete thyroxine
	Prolactin	Causes milk secretion
Posterior Pituitary	Oxytocin	Causes uterus to contract
	Vasopressin	Causes kidneys to reabsorb water
Thyroid	Thyroid Hormone	Regulates metabolic rate
	Calcitonin	Lowers blood calcium levels
Parathyroid	Parathyroid Hormone	Increases blood calcium levels
Adrenal Cortex	Aldosterone	Increases Na^+ and H_2O reabsorption in kidneys
Adrenal Medulla	Epinephrine Norepinephrine	Increase blood glucose levels and heart rate
Pancreas	Insulin	Decreases blood sugar levels
	Glucagon	Increases blood sugar levels
Ovaries	Estrogen	Promotes female secondary sex characteristics and thickens endometrial lining
	Progesterone	Maintains endometrial lining
Testes	Testosterone	Promotes male secondary sex characteristics and spermatogenesis

Summary of Pituitary Gland Hormones

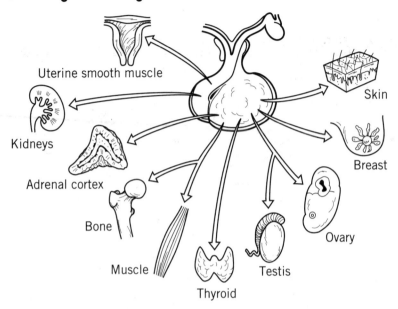

Steroid vs. Peptide Hormones

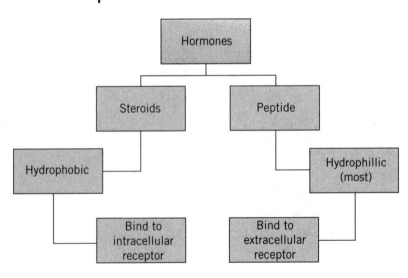

Steroid Hormone Signaling Example

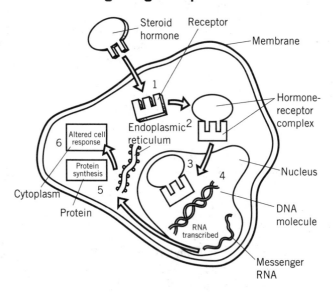

Peptide Hormone Signaling Example

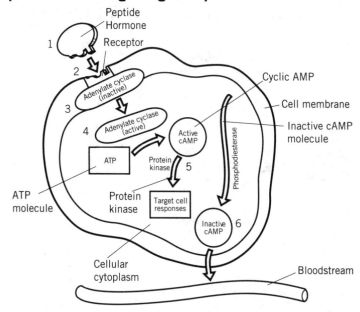

Secondary Messengers ❗

In signal transduction pathways, secondary messengers are often essential to the cascade. Secondary messengers like cyclic AMP, cyclic GMP, calcium ions, and IP3 are part of many signal transduction pathways.

Feedback ❗

The body uses internal and external cues to determine what needs to be expressed/produced/released/inhibited/etc.

Negative Feedback ❗

This is the most common type of feedback. Something at the end of a process signals that something at the beginning of the process should be turned off or inhibited.

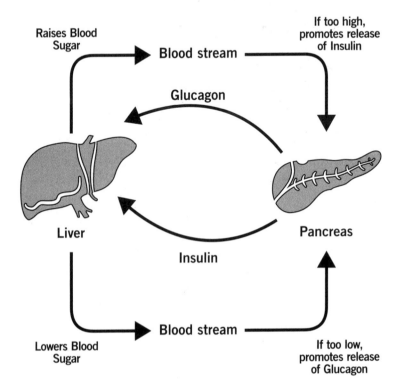

Positive Feedback—Helps amplify a signal and make it stronger. In pregnancy, positive feedback helps induce contractions and labor.

Embryonic Development ❗

The embryo develops due to specifically timed expression of genes leading to cell specialization.

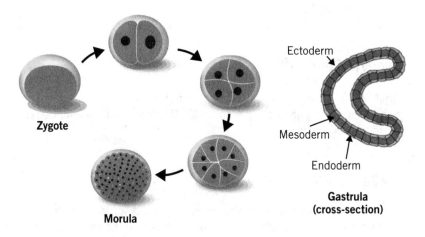

Zygote

Morula

Ectoderm

Mesoderm

Endoderm

Gastrula (cross-section)

Induction 💬

A single-celled zygote divides and different sections become specialized, as genes are turned on and off at specific times.

Some cells called organizers release special molecules called morphogens that spread from cell to cell and create a gradient. This gradient provides a "layout" for the embryo so that some parts express different development genes than other parts, and thus turn into different types of cells.

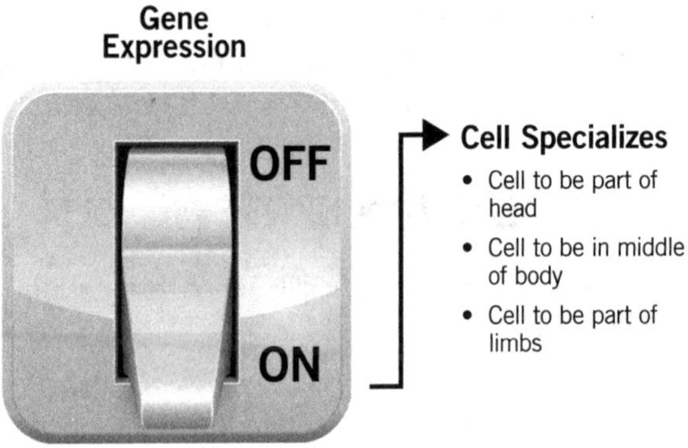

Gene Expression

OFF → **Cell Specializes**
- Cell to be part of head
- Cell to be in middle of body
- Cell to be part of limbs

ON

Homeotic Genes 💬

Some of the most important developmental genes are called homeotic genes. Homeotic genes are genes that influence the anatomical development of an organism.

Hox Genes 💬

Hox genes are a type of homeotic gene. They code for transcription factors that are extremely important in development of proper body parts. Deletion or duplication of these genes has a huge impact on the developing embryo.

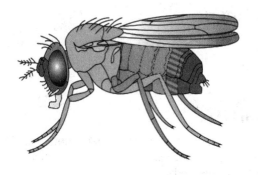

Antennapedia

lab pb   Dfd Scr Antp

Lab: Labial
Pb: Proboscipedia
Dfd: Deformado
Scr: Comba sexual reducida
Antp: Antennapedia

Bithorax

Ubx   Abd-A   Abd-B

Ubx: Ultrabithorax
Abd-A: Abdominal A
Abd-B: Abdominal B

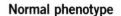

Normal phenotype

Bithorax phenotype

When the HOX genes for bithorax are mutated you get a duplicate of the second thorax segment.

Differentiation 💬

All cells share the same genome, but genes are switched on and off using transcription factors.

Stem Cells 💬

As genes are switched on and off by transcription factors, stem cells differentiate into different cell types. These specialized cells have different functions in the body.

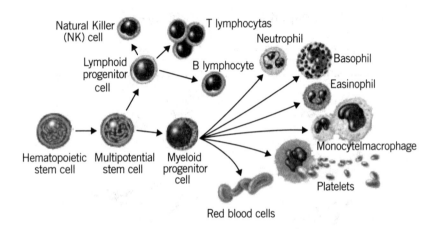

Apoptosis 🛈

Apoptosis, controlled cell death, plays a major role in normal cell development. In development, apoptosis removes the tissue between fingers and toes. In frogs, apoptosis removes the tail of tadpole during development into and adult frog.

A. Removal of Tissues

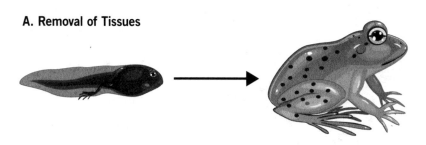

B. Organ sculpting

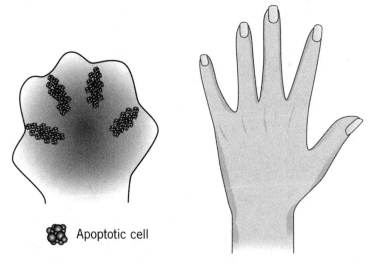

Apoptotic cell

CHAPTER 9

Behavior and Ecology

The environment is a complex balance of living and nonliving things. By studying their many interactions, the importance of this balance can be better understood and predictions can be made about the planet.

Ecosystems, Communities, and Populations ❶

Organisms can be organized into a population containing the species they belong to, the community of all living things in a region, and the ecosystem, which also includes the non-living things that they would interact with.

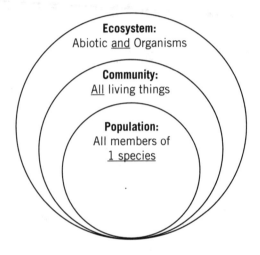

Ecosystem:
Abiotic <u>and</u> Organisms

Community:
<u>All</u> living things

Population:
All members of
<u>1 species</u>

Ecosystems ❗

An ecosystem is a region containing living organisms (biotic factors) and the non-living components (abiotic factors) of the environment that they interact with. An example could be a lake within a forest; the atmosphere surrounding the lake contains abiotic factors and all the organisms that inhabit this area are biotic factors.

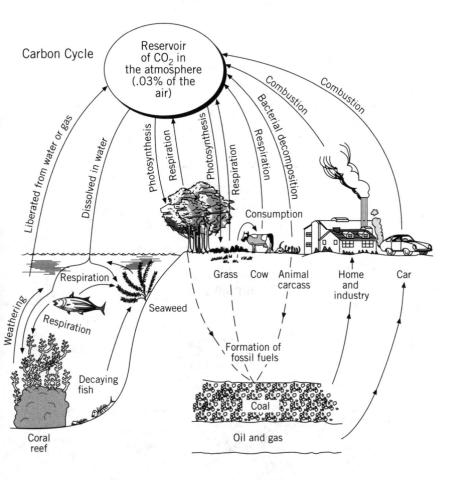

The flow of carbon throughout the ecosystem illustrates the interaction between abiotic and biotic factors.

Community 🛑

A community is made of the living inhabitants of an ecosystem. The members are either producers, consumers, or decomposers. A community is measured by how many species are in it and how many members of the species are present.

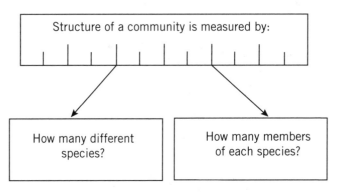

Niche 🛑

A niche is the role that an organism has in an ecosystem.

Niche = Role in an ecosystem

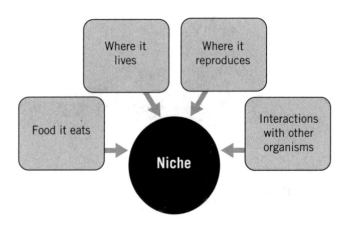

Food Chain 🛈

Food chains show how energy and nutrition are passed from one organism to another. The arrows point from the prey (eaten) to the predator (eater), thus they show the direction that energy is moving.

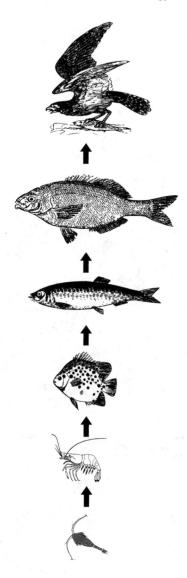

Food Webs ⚠

Since organisms typically have more than one nutrition source, a more accurate representation of predation interactions is a food web. Food webs demonstrate all the interactions of prey with predators.

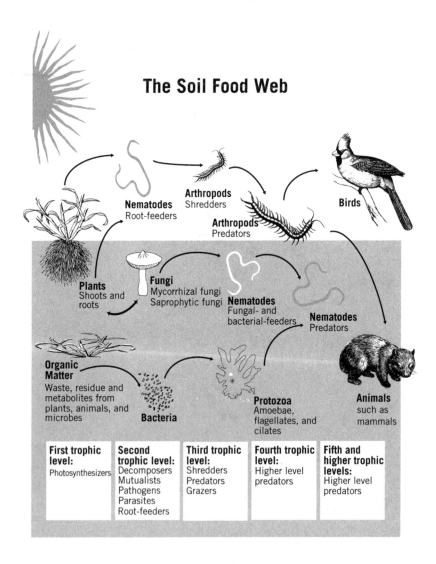

The Soil Food Web

First trophic level:	Second trophic level:	Third trophic level:	Fourth trophic level:	Fifth and higher trophic levels:
Photosynthesizers	Decomposers Mutualists Pathogens Parasites Root-feeders	Shredders Predators Grazers	Higher level predators	Higher level predators

Trophic Level Pyramid ⚠

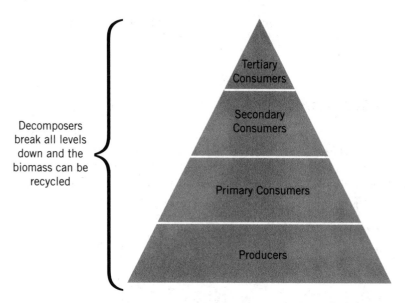

Decomposers break all levels down and the biomass can be recycled

Tertiary Consumers

Secondary Consumers

Primary Consumers

Producers

- **Producers**

  Producers are organisms that utilize photosynthesis or chemosynthesis to convert energy
  into organic substances that other organisms
  can use. Primary productivity is the rate at
  which this energy conversion occurs. They are
  often called the first trophic level.

- **Consumers**

  Consumers ingest and obtain energy as well as
  organic matter from producers or other organisms. Primary consumers (e.g. nematodes)
  ingest producers. Secondary consumers
  (e.g. arthropods) ingest primary consumers.
  Tertiary (or higher) consumers (birds) ingest
  secondary consumers. These comprise the
  second to fifth (and higher) trophic levels.
  Energy moves from lower to high trophic
  levels.

- **Decomposers**

  Break down dead members of the community into simple forms of matter. Put simply, they are the recyclers.

The 10% Rule ❗

When a plant is consumed, only 10% of its energy is retained by the primary consumer (as part of the animal's body) to be consumed by the secondary consumer. The remaining 90% of the plant energy is used up in daily functions such as growth and metabolism. This applies to every trophic level. Only 10% of the energy in one level can be passed to the next level.

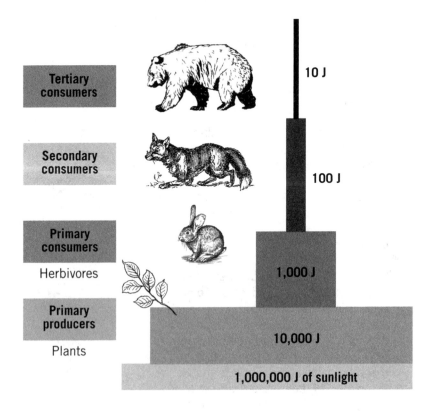

Populations and Growth ❗

Populations are all members of one species within a community. The growth of a population can be studied to learn more about the health of the population and to make predictions about the future.

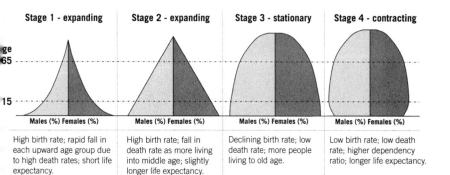

Stage 1 - expanding	Stage 2 - expanding	Stage 3 - stationary	Stage 4 - contracting
High birth rate; rapid fall in each upward age group due to high death rates; short life expectancy.	High birth rate; fall in death rate as more living into middle age; slightly longer life expectancy.	Declining birth rate; low death rate; more people living to old age.	Low birth rate; low death rate; higher dependency ratio; longer life expectancy.

Mathematically, we use the following equation:

Population Growth

$$\frac{dN}{dt} = B - D$$

Variable	Definition
dN	Change in population size
dt	Change in time
$\dfrac{dN}{dt}$	Population growth rate
B	Birth Rate
D	Death Rate

Exponential Growth

If there are no restrictions on the growth of a population, it will show exponential growth. This J-shaped curve indicates that the population rises rapidly because no predators, disease, space, or resource limitations are inhibiting the growth. When bacteria start growing in a dish, they show exponential growth.

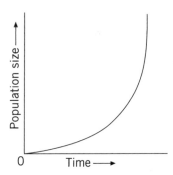

Mathematically, we use the following equation:

Exponential Growth

$$\frac{dN}{dt} = r_{max}N$$

Variable	Definition
dN	Change in population size
dt	Change in time
$\dfrac{dN}{dt}$	Population growth rate
r_{max}	Maximum per capita growth rate of population
N	Population size

Some bacteria can reproduce in less than 10 minutes.

Logistic Growth ❗

Most populations are limited by their ecosystem. There are only so many organisms that can live in a given area. This is called the carrying capacity for that population. It results in an S-shaped curve because the population will first show exponential growth until it levels out at the carrying capacity.

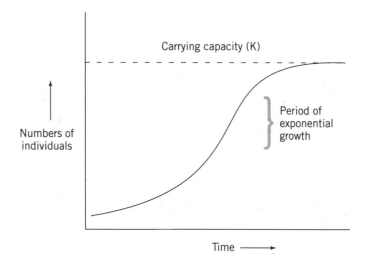

Logistic Growth

$$\frac{dN}{dt} = r_{max}N\left(\frac{K - N}{K}\right)$$

Variable	Definition
dN	Change in population size
dt	Change in time
$\dfrac{dN}{dt}$	Population growth rate
r_{max}	Maximum per capita growth rate of population
N	Population size
K	Carrying Capacity

Density-Dependent Limiting and Density-Independent Limiting Factors 🛈

- Density-dependent factors are those that decrease the population numbers in a way that is related to the size of the population. If the population is higher, the factors will have a greater effect.

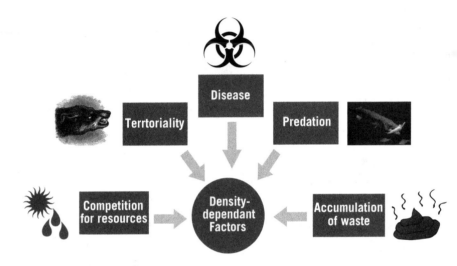

- Density-independent factors decrease a population's numbers whether the population is dense or not. These include natural disasters and toxins.

Selection Types ❗

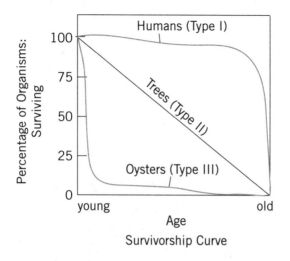

Humans (Type I)

Trees (Type II)

Oysters (Type III)

Percentage of Organisms: Surviving

100 — 75 — 50 — 25 — 0

young   Age   old

Survivorship Curve

K vs. r-Selected Species

Variable	Selection Type	
	r	K
Environment	Unstable	Stable
Organism Size	Small	Large
Energy Requirements for reproduction	Low	High
# of offspring	Many	Few
Timing of Maturation	Early	Late (requires parental care)
Life Expectancy	Short	Long
Lifetime reproductive events	One	Multiple
Survivorship Curve	Type III	Type I or II

 Ask Yourself...

Beluga whales tend to have a single offspring every two to three years and have an average lifespan of 30–35 years. What selection type do they utilize?

Behavior 🔵

Organisms respond to changes in their environment through behavioral and physiological mechanisms. These responses are often vital to reproduction and survival.

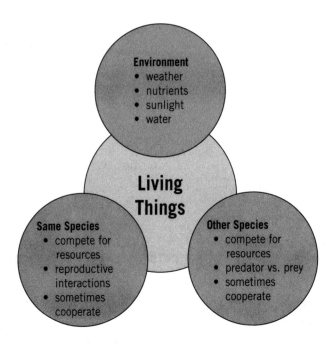

Innate and Learned Behaviors 💬

Some behaviors are innate and animals are born with the ability. Other behaviors are learned through interactions with the environment and other organisms.

Innate Behavior "born with it"	Learned Behavior "taught by someone"
• Circadian rhythms (natural 24 hr cycles)	• Avoidance behaviors (electric fences, poisons, traps)
• Reproductive displays	• Hunting tactics
• Pheromone responses	

Examples of Behavior ❗

Living things use visual, audible, tactile, electrical, and chemical signals to indicate dominance, find food, establish territory, and ensure reproductive success. Many of these behaviors have been naturally selected over time since they lead to increased survival and reproductive success.

Responses to Environmental Cues

Phototropism: Plants grow toward the sun

Photoperiodism: Plant cycles of growth change with seasons/sunlight

Hiberation: Period of inactivity due to seasonal changes affecting resources

Migration: Movement due to seasonal changes affecting resources

Chemotaxis: Movement of bacteria toward or away from a chemical

Shivering/Sweating: Physical response to temperature

Responses to Intraspecies Cues

Reproductive Communication:
Estivation: cycles of fertility
Courtship: Mating dances, displays of dominance, and calls to attract a mate

Resources Communication:
Sharing: Bee dances telling location of nectar
Non-sharing: Displays of dominance and territory marking

Protection Communication:
Packs, Herds, Schools, Flocks, Colonies, Swarms: Coordinated group movement to deter predation
Predator warning: Sounding an alarm to warn of nearby predators
Protection of young: Familial and community protection of offspring

Responses to Interspecies Cues

Predator-prey Response: Predators hunt/chase and prey defend/flee

Symbiosis: Differing species live and grow in response to each other (shown in figure on page 214)

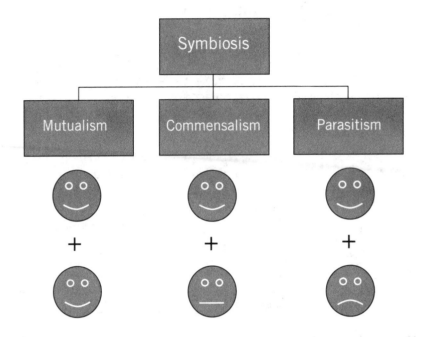

Disruptions to Ecosystems

Since ecosystems are full of complex interactions, many things can disrupt their balance. Additions or subtractions of species or resources can affect an entire ecosystem.

The Importance of Biodiversity ❗

A diverse community or population is stronger and can withstand changes better than a uniform group. This is because there are not many threats that can affect all things equally. A disease might wipe out some individuals, but not all of them. A frost might kill some plants, but not all of them. Diversity is important for a strong ecosystem.

Biodiversity = Ecosystem Strength

Invasive Species ❗

Invasive species are species that are transplanted into a new ecosystem and make a large impact. This impact is often because they have no natural enemies in the new location or because they can occupy an unused niche. This causes their population to thrive and the balance of species in the ecosystem is destroyed as the invasive species consumes resources needed by the other species. The arrival of an invasive species often occurs due to human travel. Zebra mussels, Asian carp, and kudzu are examples of invasive species.

Keystone Species ❗

A keystone species is a member of a community that is particularly important to the balance of the ecosystem. Without the keystone species, the balance of producers, consumers, and resources would be strongly affected. Of course, all members are important in an ecosystem, but keystone species are irreplaceable. Examples of keystone species include wolves, sea otters, and sea stars.

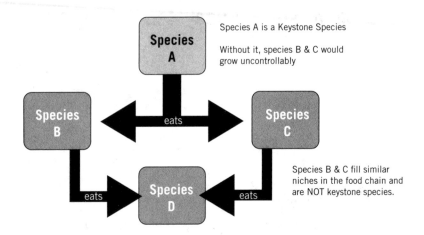

Species A is a Keystone Species

Without it, species B & C would grow uncontrollably

Species B & C fill similar niches in the food chain and are NOT keystone species.

Ask Yourself...

What would happen to an ecosystem if a keystone species like wolves was removed from an ecosystem?

Human Impact 📛

Humans often have a large effect on the environment. The following are all ways that humans can change an ecosystem. Since humans have grown in population, their effect has been magnified. This can lead to extinction.

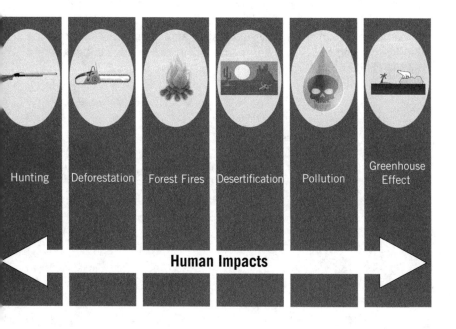

Hunting   Deforestation   Forest Fires   Desertification   Pollution   Greenhouse Effect

Human Impacts

Natural Disasters 📛

When natural disasters occur, it can be devastating to an ecosystem. Populations of living things, their resources, and their habitats can be wiped out by floods, earthquakes, fires, volcanoes, droughts, heat waves, freezing, tornadoes, tsunamis, and so on.

Ecological Succession ❗

There is a predictive pattern of growth that often occurs as ecosystems change and develop over time. This is because certain species require other species in order to thrive. In primary succession, a new habitat is colonized by species that can survive on their own in a harsh, barren environment. As the environment becomes populated and no longer barren, other species begin to grow. Eventually, a complex, rich ecosystem can develop. In secondary succession, an ecosystem is destroyed, often by fire, and the process of regrowth follows a predictable pattern until the ecosystem is restored.

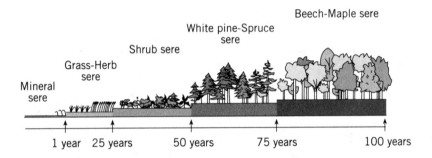

Beech-Maple sere

White pine-Spruce sere

Shrub sere

Grass-Herb sere

Mineral sere

1 year   25 years   50 years   75 years   100 years

Sometimes fires are prescribed for a given ecosystem and are intentionally lit to help manage a forest and make it stronger. Just like a prescription humans would take, the fire acts as a way to prime an ecosystem for new growth.

CHAPTER 10

Quantitative Biology and Biostatistics

As biological phenomena are observed, patterns often emerge. How do we know what is significant or valid? How do we know if what we are seeing is meaningful? Gathering and thoroughly analyzing that data allows us to determine if the patterns are valid or significant.

Data Representations

Things to Look For

- Labels
- Trends (increasing, decreasing, oscillating)
- Direct vs. indirect/inverse relationships
- Extreme values
- Statistics

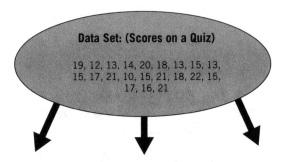

Data Set: (Scores on a Quiz)

19, 12, 13, 14, 20, 18, 13, 15, 13, 15, 17, 21, 10, 15, 21, 18, 22, 15, 17, 16, 21

Mean :

$$\frac{Sum\ of\ all\ values}{Number\ of\ values}$$

$$\frac{345}{21} = 16.43$$

Median:

When all numbers are arranged this is the middle number, or average, of the two middle numbers.

10, 12, 13, 13, 13, 14, 15, 15, 15, 15, 16, 17, 17, 18, 18, 19, 20, 21, 21, 21, 22

16 is the 11th of 21 numbers and is therefore the median

Mode:

The most common number in a data set.

15 appears 4 times so it is the mode

Direct versus inverse:

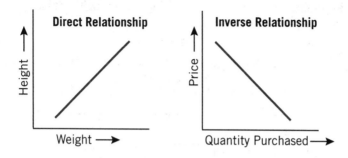

Pie Graphs 💢

Good for showing percentages of a whole.

E.coli Colony Growth on Plates after 24 hours

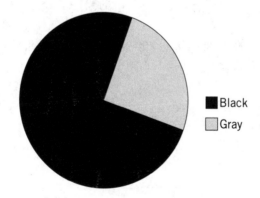

Histogram ❗

A representation of how many of a thing are in each category (often these categories are ranges of numbers).

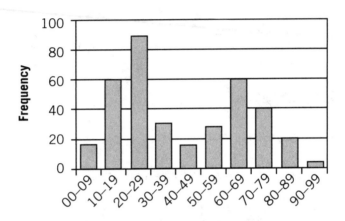

Pine Needles per Branch of *Pinus strobus* in Northern New York

Number of Pine Needles Per Branch

Bar Graph ❶

These are good for comparing levels for different categories. Since only one value is given per category, the plotted value is often an average of the data for that category.

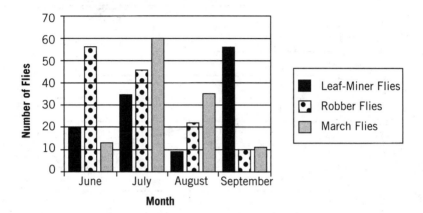

Fly Populations in Northern Maine

Line Graph ❶

This is good for comparing data when the categories are sequential (numerical, chronological, etc.). This makes it easy to identify the trend over the sequentially changing categories. Standard error bars show the variability in the data.

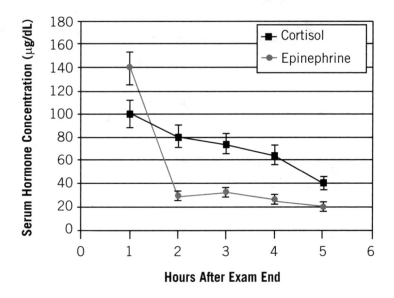

Box and Whisker ❗

A plot good for showing off sets of nonparametric numerical data in different categories. Unlike a bar graph, the variability of each categories' data is reflected. The top and bottom "whiskers" represent the most extreme values. The central box represents the middle 50% of values, and the horizontal line in the box represents the median.

Grass and Cactus Weights in the Arizona Desert

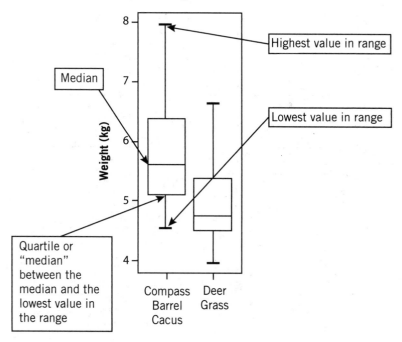

Scatterplot ❗

This is necessary when data is being compared and each piece of data contains two numerical values. This can also be helpful for showing whether two variables are correlated and what trendlines best match the data.

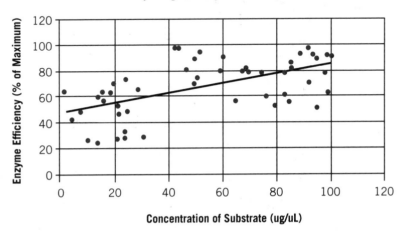

Enzyme Efficiency and Substrate Concentration for Kozmase III
at Physiological Temperature and pH

Types of Experiments

Scientific Method

The scientific method is a process that allows knowledge to be investigated. It helps to ensure that the scientific community believes you when you explain what you have observed.

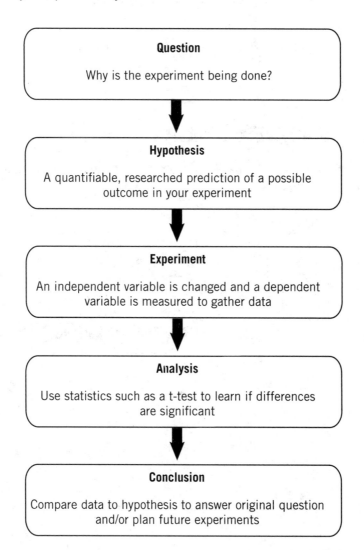

Question

Why is the experiment being done?

Hypothesis

A quantifiable, researched prediction of a possible outcome in your experiment

Experiment

An independent variable is changed and a dependent variable is measured to gather data

Analysis

Use statistics such as a t-test to learn if differences are significant

Conclusion

Compare data to hypothesis to answer original question and/or plan future experiments

Time-Course Experiments

These experiments look at how a dependent variable changes over time as an independent variable changes.

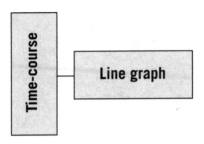

Comparative Experiments

Compare levels of a variable of interest over multiple populations, groups, or events.

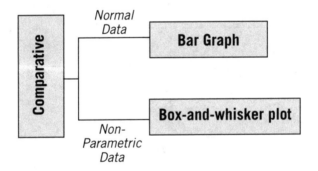

Association Experiments

Look at how two quantitative variables are correlated.

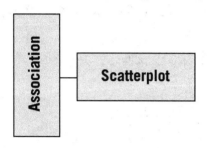

Probability ❗

The probability of an event occurring is the number of favorable cases divided by all possible cases.

The "AND" Rule ❗

If you are trying to determine the probability of independent events happening together, you MULTIPLY the probabilities.

Example: What is the probability of rolling a  ?

$$\frac{1}{6} \times \frac{1}{6} = \frac{1}{36}$$

The "EITHER/OR" Rule 🔴

If you are trying to determine the probability of EITHER one event OR the other event occurring, and the two events are <u>mutually</u> exclusive (they can't occur at the same time): you ADD the two probabilities.

Example: The probability of pulling EITHER an 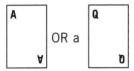 OR a

from a deck of cards $= \dfrac{4}{52} + \dfrac{4}{52} = \dfrac{8}{52}$.

However, if you are trying to determine the probability of EITHER one event OR the other event occurring, and the two events are <u>not mutually exclusive</u> (they can occur at the same time): you ADD the two probabilities and SUBTRACT the odds of them ocurring together (subtract the result from using the "AND" rule).

Example: The probability of pulling EITHER an 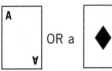 OR a

from a deck of cards $= \dfrac{4}{52} + \dfrac{13}{52} - \left(\dfrac{4}{52} \times \dfrac{13}{52} \right) = \dfrac{52}{169}$.

Statistical Significance 🔴

Statistical significance determines whether a relationship between two variables is legitimate or if it is just a chance variation.

p-values 🔴

Statistics give us confidence in results, and p-value is a measure of HOW confident we can be about our conclusion.

If a coin is flipped 100 times and it comes up as tails 80 times, someone might say that it was just a random fluke chance. Flukes happen, right? Or, is that too many tails to be a fluke?

Statistics tell us just how likely something is to be a fluke.

Scientists usually use two cutoff points: $p < 0.05$ or $p < 0.01$.

These numbers represent the probability that something would occur by random chance. Let's convert them to percentages: 5% and 1%.

With a $p < 0.05$, the data would only occur as a random fluke event 5% of the time. You could also say you are 95% confident that the relationship you see between the data is legitimate.

With $p < 0.01$, you would be 99% confident!

Standard Deviation ❗

- Standard deviation (SD) is a measure of how much variability, or spread, there is in a data set.
- A high SD means that the values are really spread out and a low SD means values are close to the mean.

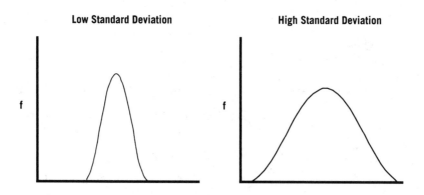

Low Standard Deviation

f

High Standard Deviation

f

Standard Error of the Mean 💬

The standard error of the mean (SEM) is the standard deviation of a data set divided by the square root of the sample size (n), which is the number of data points. The more data you collect, the smaller the standard error of the mean. It is typically shown on a graph as error bars.

$$SEM = \frac{\text{Standard Deviation}}{\sqrt{\text{sample size}}}$$

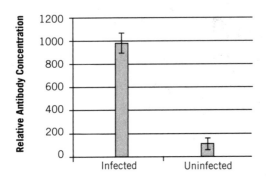

Relative Antibody Concentration with Infection

These error bars are nonoverlapping, meaning the difference between the two categories is significant.

 Ask Yourself...

If the error bars are really large, what does that mean about the data set?

Chi-square Goodness of Fit Test 💬

This test determines whether the data fits the mathematical outcome of a particular hypothesis.

- If the actual data is similar to this particular outcome, then maybe the hypothesis is true.
- If the actual data is sufficiently different from that particular outcome, then we know that hypothesis can be ruled out.

The hypothesis that is being tested is the null hypothesis. Sometimes the null hypothesis is the same hypothesis that you might suspect is true, but sometimes it might be a hypothesis you suspect is NOT true and you want to rule it out.

The formula is:

$$\chi^2 = \Sigma \frac{(o - e)^2}{e}$$

o = Observed data (the data you collected)

e = Expected data (the data you would expect if the null hypothesis was true)

Σ (sigma) = means to add everything up.

Example: A student crosses two pea plants. One he believes to be heterozygous and the other has wrinkled seeds. He knows that round (R) seeds are dominant to wrinkled (r) seeds. In the offspring, he finds 125 seeds are wrinkled and 75 are round. Did the data fit what the student should have expected?

	R	r
r	Rr	rr
r	Rr	rr

As seen in the Punnett square, if it was a heterozygote crossed with a homozygote recessive, the expected ratio of round to wrinkled seeds should be 100:100.

- The null hypothesis is that the offspring represent a cross between a heterozygote and a recessive homozygote.
- The observed were 75 round, 125 wrinkled.
- The expected are 100 round and 100 wrinkled.

Using the formula for chi-square, the following value is generated:

	o	e	o – e	(o – e)^2	(o – e)^2/e
Round	75	100	25	625	6.25
Wrinkled	125	100	25	625	6.25
Total					12.50

The χ^2 value is 12.50.

Next, determine the number of degrees of freedom (DF). This always equals the number of independent variables minus one. For this problem DF = 2 – 1 or 1 degree of freedom.

Then we compare this value to a critical value. You can choose critical values for $p < 0.05$ or $p < 0.01$. We will use $p < 0.05$ and the critical value for 1DF is 3.84.

Chi-square Table								
p	1	2	3	4	5	6	7	8
0.05	**3.84**	5.99	7.82	9.49	11.07	12.59	14.07	15.51
0.01	6.64	9.21	11.34	13.28	15.09	16.81	18.48	20.09

- If the χ^2 value is LESS than the critical value, the data is statistically similar to the null-hypothesis-expected data and the null hypothesis could be true.
- If the χ^2 value is GREATER than the critical value, the data is statistically different from the null-hypothesis-expected data and we can reject the null hypothesis.

In this case, $12.5 > 3.84$ and the null hypothesis is rejected. The data did not represent a cross of $Rr \times rr$.

CHAPTER 11

Plants

Just like animals, plants are made of cells organized into tissues. Each tissue has a specific job, and different tissues must work together to maintain homeostasis in the plant. Plants don't eat or have a digestive system, because they are photoautotrophs. As you learn about plants in this chapter, think about how they are similar to, and different from, other organisms you've learned about in this book.

Plants Cells vs. Animal Cells

Let's review what you learned about cell biology in Chapter 2. There are some characteristics of plant cells that make them structurally and functionally different than animal cells.

Plant cells have chloroplasts and perform photosynthesis.

Plant cells are often larger than animal cells.

Plants have a cell wall made of cellulose.

Plants perform cytokinesis by forming a cell plate (made of vesicles) down the middle of the dividing cell.

A large vacuole fills up to 90% of cell volume in plants.

Many genes in plants have short, and few, introns.

Animal cells are usually about 10 to 30 micrometers in size, while a plant cell can be from 10 to 100 micrometers in size.

Animal cells don't have a cell wall. Fungi have a cell wall made of chitin, eubacteria have a cell wall made of peptidoglycan, and archaebacteria have a cell wall made of chain mail proteins.

Remember!

During cytokinesis, animal cells form a cleavage furrow made of microfilaments (actin). If you need to review this, go back to Chapter 7.

Remember!

Most eukaryotic organisms are diploid.

Ask Yourself...

Vacuoles and the cell wall help a plant with support and stability. What structures do you use for the same purpose?

A vesicle is a small bubble-like structure within a cell. It is full of fluid and enclosed by a lipid bilayer. Vesicles form during exocytosis and endocytosis.

In animals, genes tend to have many long introns, making some genes really long (although the actual protein-coding sequence are not very long). In plants, introns are usually shorter, so the gene length is shorter. If you need a refresher on gene structure, go back to Chapter 4.

Plant Structure ❗

Most plants have three main vegetative organs: roots, stems, and leaves. They grow using structures called meristems.

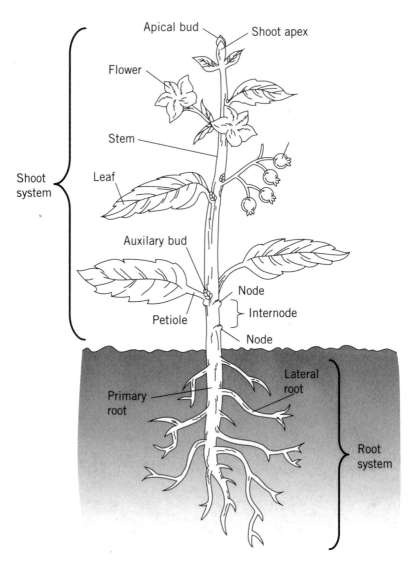

Plant Growth ❗

Meristems

- Site of plant growth
- Unspecialized cells
- Actively dividing

Primary Growth

- Increases the length/height of the plant
- Performed by apical meristems, located in the tips of roots and stems

Secondary Growth

- Increases the girth/width of the plant
- Performed by lateral meristems, located in the sides of roots and stems
- Lateral meristems produce two types of cambia
- Cambia produce layers required to make plants thicker

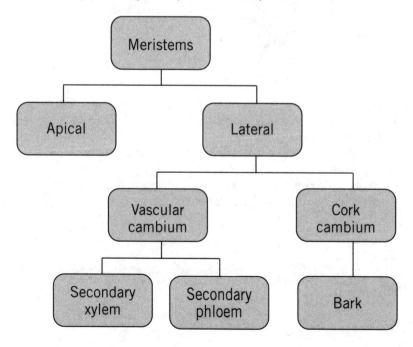

Roots 💀

- Typically below ground
- Two primary functions:
  1. Anchor the immobile plant to the ground
  2. Help with water and nutrient absorption from soil

Root Structure

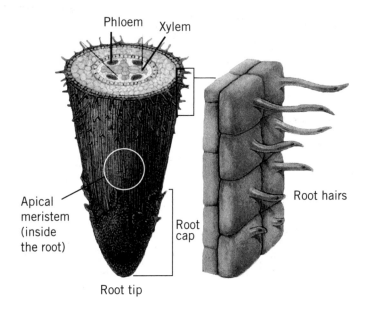

- The end of each root is called a root cap. This acts like a protective hard hat as the roots push through the soil.
- The apical meristem is where cell division occurs to push the root cap forward.
- Roots are covered with tiny little projections called root hairs, which increase the surface area of the roots.

 This is similar to how villi and microvilli increase the surface area in your small intestine. This makes sense because both of these places absorb things, and a large surface area helps this process.

Stems

Leaves attach to stems at structures called nodes. Internodes are areas of stem between nodes (or between leaf attachments). A bud is an area of new stem growth.

Stem Structure

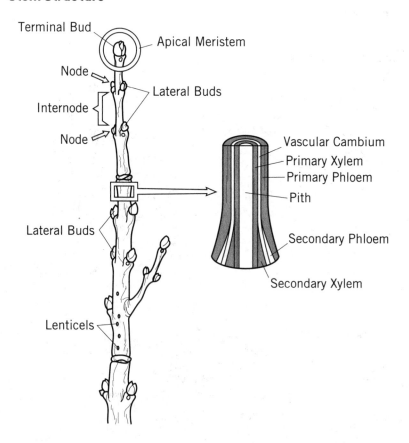

- The apical meristem is where cell division occurs to push the plant upward.
- Lenticels allow for gas exchange through the bark or outer layers of the plant.
- The pith is a low density region in the center of the stem.

Remember!

Some plants have hairs on their stems, which interfere with the feeding of some herbivores and insects. In Chapter 5 you learned that when a plant is damaged by insects or herbivores, hair density can increase in response, as a protective measure.

Leaves

Leaf Structure 💬

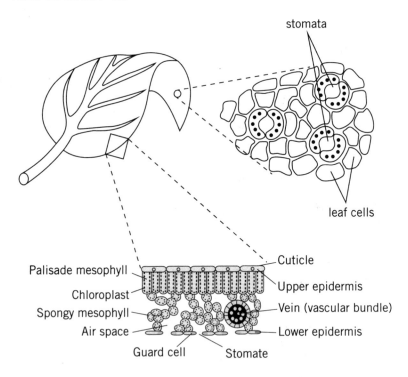

Leaf Structure	Location	Function
Epidermis	Outer-most layer	Barrier
Cuticle	Waxy covering on epidermis	Prevents water loss
Stomata	Holes or pores in the leaves	Allow exchange with air
Guard cells	Surround stomata	Control stomata
Veins	Inside the leaf	Contain vascular tissue (xylem and phloem)
Palisade mesophyll	Inside the leaf	Contains lots of chlorophyll Has high photosynthetic rates
Spongy mesophyll	Inside the leaf	Gas exchange

Remember!

Palisade mesophyll can also be called palisade parenchyma, and spongy mesophyll can also be called spongy parenchyma.

Ask Yourself...

1. The epidermis, cuticle, and stomata are crucial structures for the plant. What structures do you use for the same purposes?

2. Some of the structures you've learned about help plants fight invading pathogens: thick cell walls and waxy cuticles prevent entry of pathogens. Plants can also release antimicrobial compounds. What structures or systems do you use for the same purposes?

Vascular Tissue

Plants have three types of tissue: epidermal tissue provides protective covering, ground tissue fills the plants interior, and vascular tissue is responsible for the transport of water and nutrients. Plants that have xylem and phloem are called vascular, and plants without these structures are called avascular.

Vascular Tissue	Structure	Function
Xylem	Tracheids Vessel elements	Conduct water and minerals up from the roots
Phloem	Sieve tube elements Companion cells	Conduct nutrients (such as sugars) around the plant

Xylem Structure

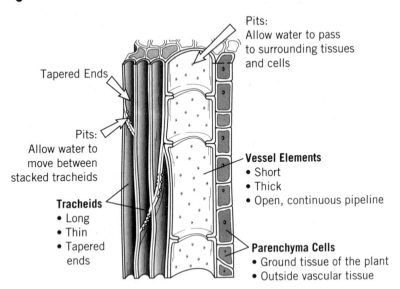

Pits:
Allow water to pass to surrounding tissues and cells

Tapered Ends

Pits:
Allow water to move between stacked tracheids

Tracheids
• Long
• Thin
• Tapered ends

Vessel Elements
• Short
• Thick
• Open, continuous pipeline

Parenchyma Cells
• Ground tissue of the plant
• Outside vascular tissue

 Remember!

Both tracheids and vessel elements are hollow, nonliving structures.

Water Transport into Roots

- When a root finds water, it absorbs it via osmosis
- Water must cross 3 different layers of cells before reaching vascular tissue: epidermis, cortex, and endodermis
  - The endodermis is a protective barrier of cells
  - The area around these cells is watertight because of a waxy strip called the Casparian strip
  - This barrier acts like a checkpoint and regulates what enters the plant and what doesn't
- After crossing the endodermis, water travels to the xylem
- Water can enter the plant via an extracellular pathway or an intracellular pathway

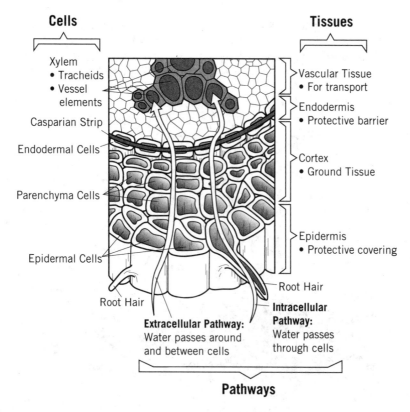

Cells

Xylem
- Tracheids
- Vessel elements

Casparian Strip

Endodermal Cells

Parenchyma Cells

Epidermal Cells

Root Hair

Tissues

Vascular Tissue
- For transport

Endodermis
- Protective barrier

Cortex
- Ground Tissue

Epidermis
- Protective covering

Root Hair

Extracellular Pathway: Water passes around and between cells

Intracellular Pathway: Water passes through cells

Pathways

How Do Plants Obtain Minerals?

- Minerals are dissolved in soil water, so they enter the plant through soil.
- Roots also actively take up minerals, which are often present at lower concentrations in the soil than they need to be in the cell.
- Remember, things naturally flow from places where there is more to places where there is less (or down a concentration gradient). If a plant has more of something than the soil, the substance will need to be actively imported.

Water Transport Up the Plant: Cohesion-Tension Theory

- Water molecules are cohesive and adhesive
- Water evaporates out of leaves (a process called transpiration)
- Creates tension that pulls water molecules up
- Water climbs the xylem because of capillary action, surface tension, and adhesion to other water molecules

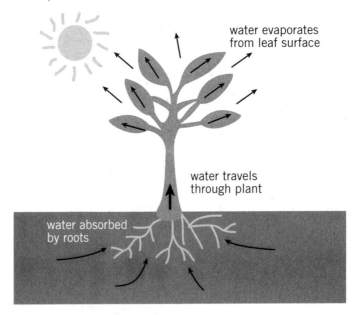

water evaporates from leaf surface

water travels through plant

water absorbed by roots

Remember!

The properties of water were reviewed in Chapter 1. If you need a refresher, check it out now.

Remember!

Think of water molecules in the xylem acting like a chain of connected particles, all climbing from roots to leaves together.

How to Plants Deal with Drought?

In response to drought, plants decrease growth, photosynthesis, and stomatal opening. This is regulated by physical and chemical signals.

Phloem Structure

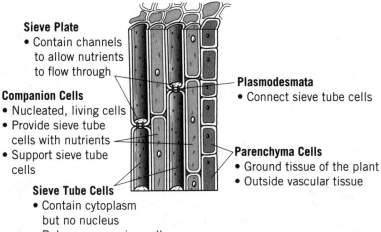

Sieve Plate
- Contain channels to allow nutrients to flow through

Companion Cells
- Nucleated, living cells
- Provide sieve tube cells with nutrients
- Support sieve tube cells

Sieve Tube Cells
- Contain cytoplasm but no nucleus
- Rely on companion cells
- Responsible for nutrient transport

Plasmodesmata
- Connect sieve tube cells

Parenchyma Cells
- Ground tissue of the plant
- Outside vascular tissue

Nutrient Transport: Translocation and Bulk Flow

- Nutrients move from source (where they are made) to sink (where they are needed)
- Companion cells actively pump nutrients into sieve tubes
- Water enters sieve tubes via osmosis
- Pressure builds up
- Nutrient solution moves to the sink
- Sugars are unloaded at the sink cell
- Water leaves the sink cell and travels back to the xylem

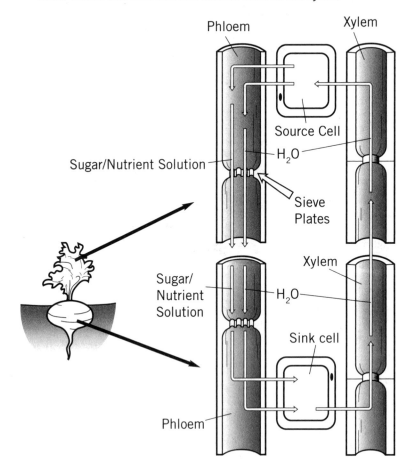

Phloem   Xylem

Source Cell

Sugar/Nutrient Solution H_2O

Sieve Plates

Xylem

Sugar/ Nutrient Solution H_2O

Sink cell

Phloem

Ask Yourself...

1. Why is it important that water goes back to the xylem after participating in nutrient flow?

2. Xylem and phloem allow vascular plants to transport water and nutrients to different parts of the organism. What structures do you use for the same purposes?

Remember!

Nutrient transport is closely coupled to water transport in plants. You can't have one without the other.

Plants require cooperation between the roots and shoots. For example, a plant would be unable to survive if its root system was too small to absorb water to replace the water lost through transpiration by the shoot.

Photosynthesis

Plants and algae are producers and perform photosynthesis: they convert light/solar energy into chemical energy. In plants, most photosynthesis occurs in leaves.

Remember!

Algae includes some bacteria (blue-green algae or cyanobacteria) and plant-like protists (other algae).

In plants, photosynthesis takes place in chloroplasts. You already reviewed chloroplast structure in Chapter 2. Review what you've learned by labelling the important organelle on the next page. You can check your work in the footnote on page 253.

Nutrients move from source to sink but since most photosynthesis occurs in leaves, this is mostly from the leaves down to the rest of the plant. A ridiculous but also awesome way to remember transport directions in plants is the phrase "Xylem to the sky-lem, phloem to the low-em."

Chloroplast Structure ❗

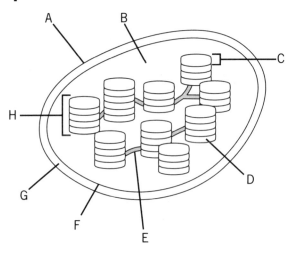

 Ask Yourself...

1. Where else have you heard the word "lumen"? Think about both cell biology and human physiology topics.
2. Which other organelles have two membranes?

Photosynthesis occurs in two steps: light-dependent reactions and light-independent reactions.

Photosynthetic Reactions	Location	Purpose	Reactant(s)	Product(s)
	Light-dependent	Thylakoid membrane	Convert light energy to chemical energy (ATP and NADPH)	Sunlight H_2O	O_2 ATP NADPH
	Light-independent	Stroma	Use ATP and NADPH to build organic molecules from CO_2	CO_2 ATP NADPH	Carbohydrates

Photosynthesis Key Points ❗

Photosynthesis is an example of carbon fixation because carbon dioxide is incorporated into complex organic molecules.

The light-dependent reactions produce the chemical energy needed by the light-independent reactions to produce organic molecules.

The net equation for photosynthesis is

$$6\ CO_2 + 6\ H_2O + energy \longrightarrow 6\ O_2 + C_6H_{12}O_6$$

This equation is essentially the chemical reverse of cellular respiration.

 Ask Yourself...

$C_6H_{12}O_6$ is the molecular formula for which molecule(s)?

Check Your Work: Chloroplast Structure: A = outer membrane, B = stroma (aqueous fluid), C = thylakoid, D = lumen (inside of thylakoid), E = lamella, F = inner membrane, G = intermembrane space, H = granum (stack of thylakoids)

Light Reactions ❗

Light-Dependent Reactions Key Points

H_2O +
Sunlight →
O_2 + Energy

Depend on light, or solar, energy

Light energy is used to make ATP from ADP and a phosphate (photophosphorylation)

Photosystems consist of chlorophyll pigment molecules

Photosynthetic electron transport chain is very similar to the electron transport chain used in cellular respiration

 Ask Yourself...

Does photophosphorylation occur in cell respiration? Do animals perform photophosphorylation?

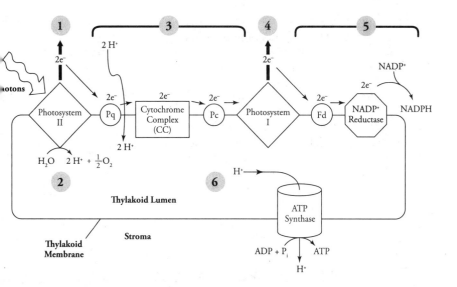

Photosynthesis: Light-Dependent Reactions 💀

1. Chlorophyll in photosystem II absorbs light
   Energy is passed to the reaction center (P680)
   Electrons in P680 are excited to a higher energy level

2. Water is split into H^+ ions and an oxygen atom (this is called photolysis) to replace the electrons lost from chlorophyll
   Two oxygen atoms combine and are released as O_2

3. Excited electrons are passed along an electron transport chain (ETC)
   Photosystem II → Plastoquinone (Pq) → Cytochrome complex (CC) → Plastocyanin (Pc)
   H^+ ions are transported from the stroma into the thylakoid lumen during the ETC

4. Chlorophyll in photosystem I absorbs light
   Energy is passed to the reaction center (P700)
   Electrons in P700 are excited to a higher energy level

5. Excited electrons continue along an ETC
    Photosystem I → Ferredoxin (Fd) → NADPH
    NADP+ reductase transfers a pair of electrons and two protons to
    NADP+, reducing it to NADPH

6. A proton gradient forms due to pumped H+ by the ETC and H+
    from the splitting of water
    Proton gradient is used to drive ATP production
    Protons diffuse down their concentration gradient (thylakoid
    lumen to stroma) through ATP synthase; this couples the move-
    ment of protons to the phosphorylation of ADP

Ask Yourself...

Does photolysis occur in cell respiration? Do animals perform
photolysis?

Remember!

There are several types of chlorophyll and light-trapping molecules
used in photosynthesis, but the predominant one in plants is chlorophyll a.

Photosystem I and photosystem II were numbered in order of their discovery, not the order
they are used in photosynthesis.

Energy Transfer in the Light Reactions ❗

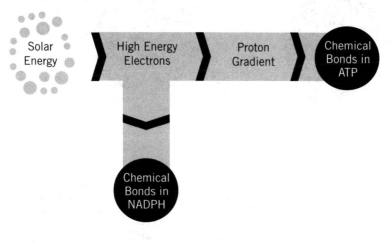

Solar Energy → High Energy Electrons → Proton Gradient → Chemical Bonds in ATP

High Energy Electrons → Chemical Bonds in NADPH

ATP synthase, NADP⁺ reductase, and all the electron transport chain machinery are located in the thylakoid membrane.

Both ATP and NADPH are made in the stroma.

 Remember!

Some bacteria perform photosynthesis but do not have membrane-bound organelles. In these cells, photosynthetic machinery is found in the plasma membrane.

 Ask Yourself...

1. Chlorophyll is the source of electrons in the photosynthetic electron transport chain. Where do the electrons come from in the cell respiration electron transport chain?

2. NADP⁺ is the terminal electron acceptor in the photosynthetic electron transport chain. What is the terminal electron acceptor in the cell respiration electron transport chain?

 NADP⁺ stands for nicotinamide adenine dinucleotide phosphate. This large molecule is similar to NAD⁺ (nicotinamide adenine dinucleotide), which you learned about in Chapter 3. Both molecules carry energy and electrons around the cell when they are reduced (to NADH and NADPH).

PSI and PSII each absorb different wavelengths of light.

PSII has a reaction center called P680, because it has a maximum absorption at a wavelength of 680 nanometers

PSI has a reaction center called P700, because it has a maximum absorption at a wavelength of 700 nanometers

3 different graphs show how pigments and photosynthesis depend on wavelength of light

An absorption spectrum shows how well a certain pigment absorbs electro-magnetic radiation.

An emission spectrum is the opposite; it gives information on which wavelengths are emitted by a pigment.

An action spectrum shows the rate of photosynthesis at the different wavelengths of visible light

Absorption Spectrum

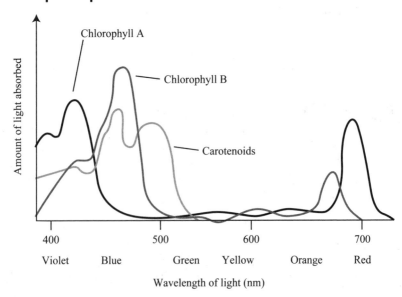

Chlorophyll A

Chlorophyll B

Carotenoids

Amount of light absorbed

400 — Violet
Blue
500 — Green
Yellow
600 — Orange
700 — Red

Wavelength of light (nm)

Ask Yourself...

How does the absorption spectrum explain why most plants are green in color?

Action Spectrum

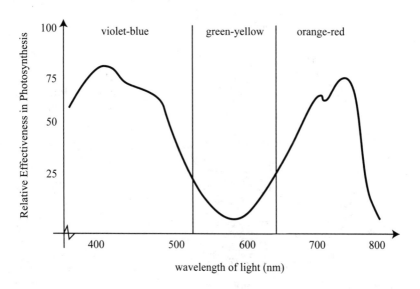

 Ask Yourself...

Why does the action spectrum have the same general shape as the absorption spectrum?

Most plants perform the light reactions in the way we just summarized, and this electron flow is called linear, or noncyclic. However, some plants (such as C_4 plants—more on these later) perform cyclic electron flow instead.

Cyclic Light Reactions 💬

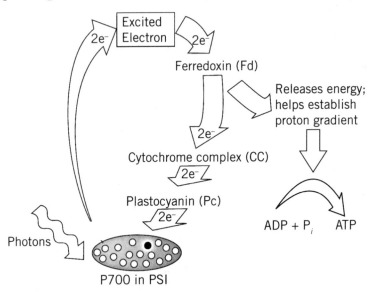

Ask Yourself...

Is the electron flow in cell respiration linear or cyclic?

	Noncyclic Light Reactions	Cyclic Light Reactions
Uses PSII	✔	✗
Uses PSI	✔	✔
Photolysis	✔	✗
Generates ATP	✔	✔
Generates NADPH	✔	✗
Electron Flow	PSII → Pq → CC → Pc ↓ NADPH ← Fd ← PSI	PSI → Fd → CC → Pc → PSI

Dark Reactions

Light-Independent Reactions Key Points ❗

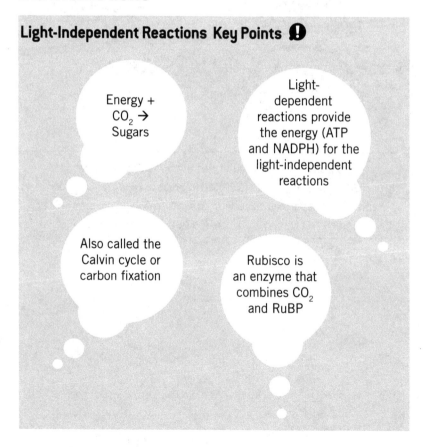

Energy + CO_2 → Sugars

Light-dependent reactions provide the energy (ATP and NADPH) for the light-independent reactions

Also called the Calvin cycle or carbon fixation

Rubisco is an enzyme that combines CO_2 and RuBP

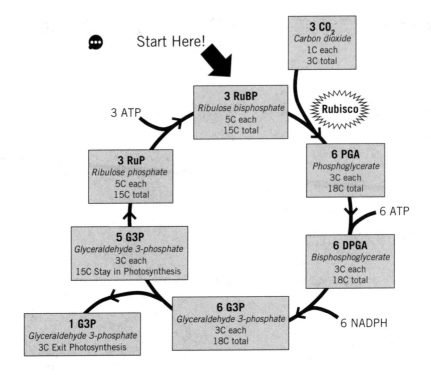

Start Here!

3 CO$_2$
Carbon dioxide
1C each
3C total

3 RuBP
Ribulose bisphosphate
5C each
15C total

3 ATP

Rubisco

6 PGA
Phosphoglycerate
3C each
18C total

3 RuP
Ribulose phosphate
5C each
15C total

6 ATP

6 DPGA
Bisphosphoglycerate
3C each
18C total

5 G3P
Glyceraldehyde 3-phosphate
3C each
15C Stay in Photosynthesis

6 G3P
Glyceraldehyde 3-phosphate
3C each
18C total

6 NADPH

1 G3P
Glyceraldehyde 3-phosphate
3C Exit Photosynthesis

The actual product of photosynthesis is a three-carbon sugar called glyceraldehyde 3-phosphate (G3P). G3P can be used to produce glucose and other organic molecules. Since it contains 3 carbon atoms, this method of producing glucose is called the C$_3$ pathway.

 Remember!

Rubisco catalyzes the carbon fixation step of photosynthesis.

 Plants breathe out oxygen and breathe in CO$_2$, the opposite of what we do.

Most scientists agree that Rubisco is probably the most abundant protein in the world, with actin taking second place.

ASAP Biology

Some Like It Hot 💬

Plants that live in hot climates run into a problem...conserving water can result in a wasteful process called photorespiration.

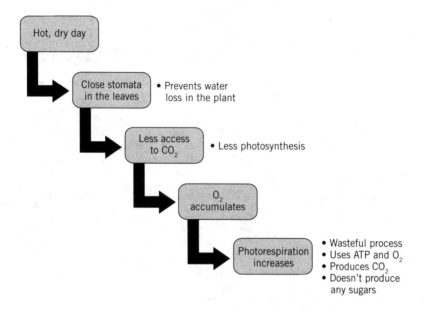

- **Hot, dry day**

- **Close stomata in the leaves**
  - Prevents water loss in the plant

- **Less access to CO_2**
  - Less photosynthesis

- **O_2 accumulates**

- **Photorespiration increases**
  - Wasteful process
  - Uses ATP and O_2
  - Produces CO_2
  - Doesn't produce any sugars

Evolution of 2 Solutions ⚠️

Plants that live in hot climates have evolved two different ways around this: CAM and C_4 photosynthesis.

C₃ Plants	CAM Plants	C₄ Plants

- Perform the light-dependent and light-independent reactions at the same time and in the same place
- Mostly linear electron flow in the light reactions
- Carbon is fixed into a 3-carbon molecule (G3P)
- Examples: potatoes (above), beans, rice, wheat

- Night: Open stomata and incorporate CO_2 into organic acids
- Day: close stomata, release CO_2 from organic acids, run light reactions
- Mostly linear electron flow in the light reactions
- Carbon is fixed into 4-carbon organic acids
- Examples: pineapple plant (above), cacti

- Slightly different leaf anatomy
- Perform CO_2 fixation in a different part of the leaf than the rest of the light-independent reactions
- Mostly cyclic electron flow in the light reactions
- Carbon is fixed into 4-carbon organic acids
- Examples: sugarcane (above), corn

 Remember!

C_3 and C_4 refer to the number of carbons *initially* involved in making sugar. Both pathways ultimately use the Calvin cycle to produce carbohydrates such as glucose.

 Ask Yourself...

Do CAM plants spatially or temporally separate the light and dark reactions? What about C_4 plants?

 CAM stands for "Crassulacean acid metabolism." It is named after the Crassulaceae family of succulents, which were used in the discovery of this pathway.

Summary of Photosynthesis ❗

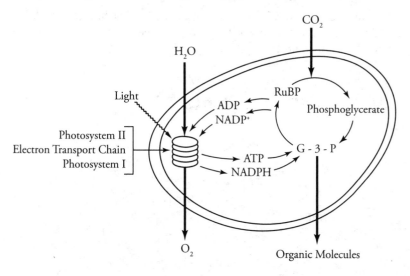

Taxonomy of Plants

There are about 300,000 species of plants on Earth. They are found in most environments on Earth, supply oxygen, and are a major food source for many animals.

Characteristics of Plants ❗

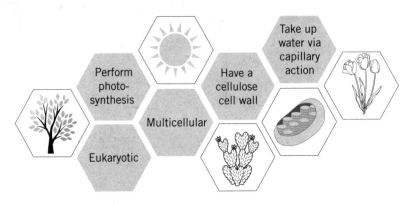

 Remember!

All cells have a plasma or cell membrane. Not all cells have a cell wall.

💬 Classifying plants is useful because it gives you a sense of how they evolved.

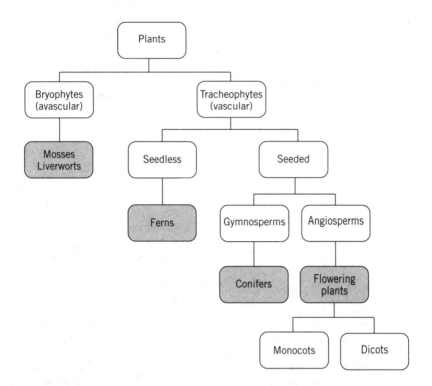

Plant	Roots, Stems, Leaves	Vascular Tissue	Seeds	Flowers	Dominant Generation*
Bryophytes	✗	✗	✗	✗	Gametophyte
Tracheophytes	✔	✔	Variable	Variable	Sporophyte
Ferns	✔	✔	✗	✗	Sporophyte
Gymnosperms	✔	✔	✔	✗	Sporophyte
Angiosperms	✔	✔	✔	✔	Sporophyte
Monocots	✔	✔	✔	✔	Sporophyte
Dicots	✔	✔	✔	✔	Sporophyte

*Alternation of generations will be reviewed in the next section

Tracheophytes

- Have roots, stems, and leaves
- Vascular: have xylem or phloem

Bryophytes

- Lack true roots, stems, and leaves
- Avascular: no xylem or phloem
- Seedless

Gymnosperms

- Woody plants such as conifers
- Perennial
- Seeds are found in cones

Angiosperms

- Flowering plants
- Seeds are enclosed in a fruit or nut
- Includes monocots and dicots

Remember!

Seedless plants (all bryophytes and some tracheophytes) need an abundant water supply for fertilization: water allows the sperm to make their way to egg cells in reproduction.

Monocots vs. Dicots 💬

A cotyledon is part of the embryo within the seed of a plant. It becomes the first leaves to form from a developing seed. You'll learn more about this in the next section. Monocots form one leaf during seed development, and dicots form two.

Floral parts in 3s   Floral parts in 4s or 5s

One cotyledon   Two cotyledons

Fruit wall fused to seed coat
Cotyledon
Endosperm

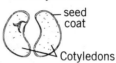

seed coat

Cotyledons

Scattered vascular bundles   Vascular bundles in circle

Xylem

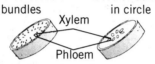

Phloem

Fibrous root system   Taproot system

Parallel venation; long, tapering blade with sheath encircling stem

Netted venation; expanded blade and petiole

Monocots

Examples: grasses, corn, wheat

Dicots

Examples: legumes (peas and beans), most flowering trees (oaks and maples).

 About 1/3 of angiosperms are monocots (70,000 species) and about 2/3 of angiosperms are dicots (170,000 species).

Angiosperms have evolved several mechanisms to reproduce sexually with members of their own species that are located far away. Beautiful flowers, sticky substances, slippery substances, wind, insects, bats, and large animals with fur can all help flowering plants mate. In this section we will review how various types of plants reproduce.

Alternation of Generations ❗

The plant life cycle alternates between two different forms. One generation of the plant is haploid (called a gametophyte) and generates gametes. The next generation is diploid (called a sporophyte) and generates spores.

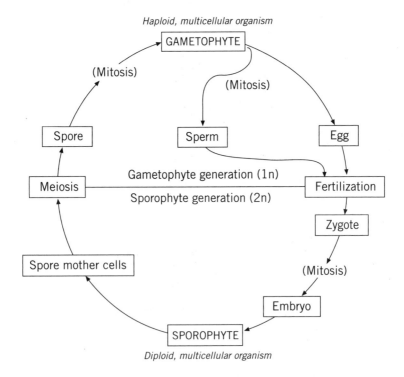

Haploid, multicellular organism
Diploid, multicellular organism

Remember!

Fertilization is when an egg and a sperm fuse to generate a zygote. This is true whether you're talking about an animal (such as a human), a plant, or a protist.

Both sporophytes and gametophytes are separate, multicellular, independent organisms, and may or may not have a similar appearance.

All plants alternate between diploid and haploid forms, but one of the two forms are usually dominant in both size and generation length.

$$2n > n$$
$$2n < n$$

Some protists and some fungi also undergo alternation of generations.

Bryophytes are gametophyte-dominant.

Tracheophytes are sporophyte-dominant.

Ask Yourself...

If humans did alternation of generations, would you have the same number of chromosomes as your parents? What about your grandparents?

Bryophyte Reproduction

Here is the life cycle of moss, a common bryophyte. Most of the mosses you see are gametophytes because that is the dominant generation. Notice from the bottom half of the diagram that the sporophyte plant grows on top of the gametophyte plant and depends on it for its survival.

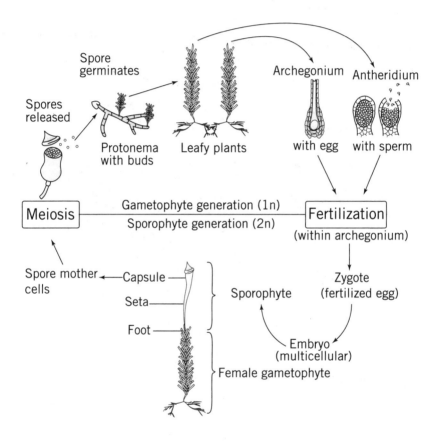

Angiosperm Reproduction 💬

Parts of a Flower

There are 4 main parts to a flower:

1. Sepals—green, leaf-like structures that cover and protect the flower
2. Petals—brightly colored to attract potential pollinators
3. Stamen—male components
4. Pistil—female components

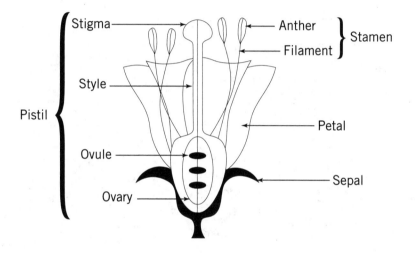

Stamen =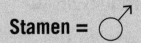

Filament:
holds up
anther

Anther:
produces
pollen grains

Pistil = ♀

Stigma:
sticky,
captures
pollen grains

Style:
tube that
connects
stigma and
ovary

Ovary:
where
fertilization
occurs

 Some plants increase the temperature of their floral parts to help attract pollinators.

Flower colors like red, pink, blue, and purple come from pigments called anthocyanins, which are in the class of chemicals called flavonoids. However, as you learned in Chapter 5, other factors such as soil pH or nutrient availability can affect flower color.

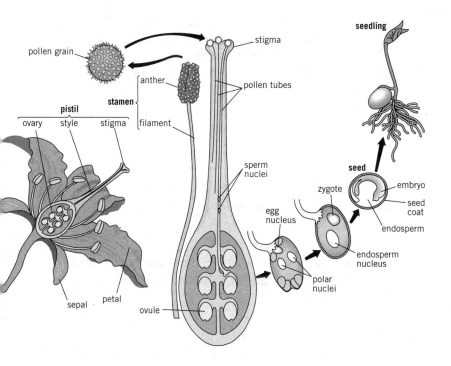

Reproduction of Flowering Plants

1. Pollen grains fall onto the stigma, which is sticky. There are many ways this can happen: they can be blown there, fall there, or be placed there by insects.
2. Once on the stigma, pollen grains germinate.
3. During germination, a pollen tube grows down through the style and connects to the ovary.
4. Two sperm (from the pollen) travel down the pollen tube and enter the ovule inside the ovary.
5. Double fertilization occurs: one sperm fertilizes the egg and the other combines with polar bodies.
6. The fertilized egg becomes the plant embryo and the polar bodies become endosperm, a food-storing tissue that surrounds the plant embryo.

7. The entire ovule, which contains the embryo and the endosperm, develops into a seed. The ovary develops into a fruit. The fruit protects the seed and helps it disperse via wind or animals.
8. The seed is released (the fruit drops, the plant is eaten, etc.) and if it finds a suitable environment, it develops into a new plant.

Seed Germination ❗

In order for a seed to germinate, it requires warm or moderate temperature, water, and rich soil. Some plants require light for later stages of development. Also remember that seeds aren't viable forever. Eventually, they will be incapable of supporting growth.

If conditions are not favorable, a seed will not germinate. Plant seed dormancy can increase the survival of a species.

Ask Yourself...

If a plant self-pollinates, is it performing sexual or asexual reproduction? How will offspring plants be related to the parental plant genetically?

Apples, pears, and oranges are all fertilized ovaries of flowering plants.

Some flowers can self-pollinate, but most have mechanisms that ensure cross-pollination (the pollen from one plant fertilizes another plant of the same species). Cross-pollination enhances genetic variability and is a form of sexual reproduction.

Ploidy of Angiosperm Reproduction

One sperm nucleus (*n*) fuses with an egg nucleus (*n*) to form a zygote
(2*n*). This zygote eventually forms a diploid sporophyte plant.
The other sperm nucleus (*n*) fuses with two polar nuclei (2*n*)
in the ovary to form the endosperm (3*n*). The endosperm does
not develop into a plant.

 Remember!

Double fertilization produces two things: a plant and food for the plant.

Flowering plants perform alternation of generations, with the sporophyte
being the dominant generation. The flower of the sporophyte produces
microspores that form male gametophytes and megaspores that form
female gametophytes. Gametophytes are haploid and found inside the
plant. The male gametophyte is inside the pollen grain and develops
into sperm cells and the pollen tube. The female gametophyte is inside
the ovary and develops into the egg.

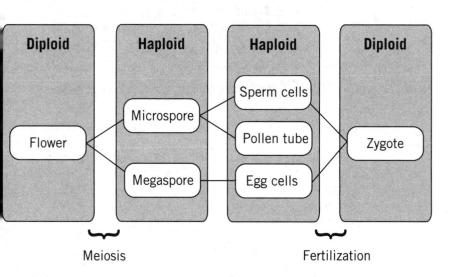

Parts of a Seed

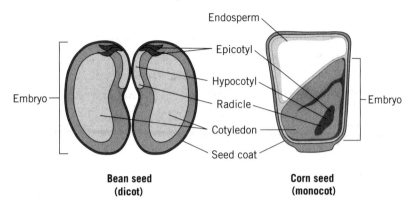

Bean seed
(dicot)

Corn seed
(monocot)

Seed Structure	Function
Seed Coat	• Protection
Endosperm	• Rich in starch and other nutrients • Nourishes the developing embryo
Cotyledons	• First embryo leaves to appear • Temporarily stores nutrients for the plant
Epicotyl	• Becomes stems and leaves
Hypocotyl	• Becomes roots
Radicle	• Embryonic root that starts root development

 Remember!

Thick cotyledons and the endosperm have the same function: food supply.

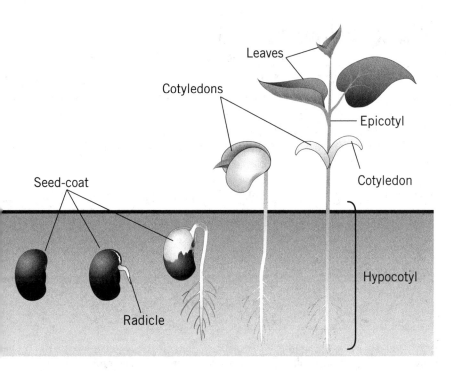

Vegetative Propagation

Many plants can reproduce asexually via vegetative propagation. This is when a new plant is formed without the production of seeds or spores. Part of the parent plant (such as a root, stem, or leaf) produces another plant. Here are some examples:

Types of Vegetative Propagation	Description	Examples
Bulbs	Short stems underground	Onions
Runners	Horizontal stems above the ground	Strawberries
Tubers	Underground stems	Potatoes
Grafting	Cut a stem and attach it to a closely related plant	Seedless oranges

Plant Behavior and Communication !

Plants are sessile organisms; they cannot move. However, that doesn't mean they don't respond to their environment and other organisms. Plants can turn and respond to the length of the day. They must also fight invading pathogens. Hormones are commonly used as signalling molecules, and cell communication occurs within and between plants.

Like animals, plants use communication and signalling pathways to control growth, development, survival, and to adapt to environmental cues. Here is a summary of plant communication:

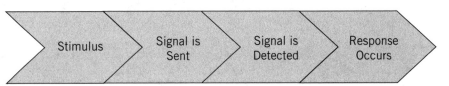

Stimulus	Signal is Sent	Signal is Detected	Response Occurs
• Light • Gravity • Touch • Length of day • Pathogen • Something happening in a nearby cell	• Hormone • Volatile chemical • Electric signal	• Usually by a receptor	Changes in: • Respiration • Chlorophyll production • Photosynthesis • Flowering • Seed germination • Wound healing • Immune response

Tropisms !

- A type of plant behavior
- Part or all of the organism turns in response to an external stimulus
- Positive tropism: when the plant goes toward the stimulus
- Negative tropism: when it goes away from the stimulus

Tropism	Stimulus	Result	
Phototropism	Light	• Changes in light leads to differential growth • Maximizes exposure of leaves to light for photosynthesis	🗩
Gravitropism	Gravity	• Stems grow up (negative gravitropism) • Roots grow down (positive gravitropism)	🗨
Thigmotropism	Touch	• Some plants can grow around a pole or trellis	🗨

 Remember!

Taxis is the *movement* of an organism in response to a stimulus, while a tropism is *turning*. Both can be positive (toward the stimulus) or negative (away from the stimulus).

Photoperiodism 💬

- A type of plant behavior
- When plants respond to the length of day or night
- Determined by phytochrome, a pigment and light receptor
- Regulates flowering and helps plants prepare for winter

Most angiosperms flower according to the amount of uninterrupted darkness. Short-day plants require a long period of darkness and bloom in late summer or fall when daylight is decreasing. Long-day plants need short periods of darkness and flower in late spring and summer when daylight is increasing. Day-neutral plants don't flower in response to daylight changes at all; they use other cues such as water or temperature.

 Ask Yourself...

Does phytochrome activation inhibit or induce flowering in short-day plants? What about in long-day plants?

Plant Hormones 💬

Gibberellins

Promote stem elongation, especially in dwarf plants

Auxins

Promote plant growth, cell elongation, fruit development, and phototropism

Cytokinins

Promote cell division and differentiation

Ethylene

Induces leaf abscission and promotes fruit ripening

Abscisic acid

Inhibits leaf abscission and promotes bud and seed dormancy

For more on plant hormones, go online and check out the ASAP Biology Supplement file posted there!

! Cell communication can occur within an organism or between organisms. For example:

- Adjacent cells communicate by direct contact
- Nearby cells or far away cells in the same plant communicate by signalling molecules
- Different plants communicate with each other using signalling molecules

---

Wounded tomatoes produce a volatile chemical as an alarm signal. This warns nearby plants and allows them to prepare a defense or immune response. A similar response has been found in trees being attacked by an herbivore.

AP BIOLOGY
EQUATIONS AND FORMULAS

Statistical Analysis and Probability

Mean

$$\bar{x} = \frac{1}{n} \sum_{i=1}^{n} x_i$$

Standard Deviation*

$$S = \sqrt{\frac{\sum (x_i - \bar{x})^2}{n-1}}$$

Standard Error of the Mean*

$$SE_{\bar{x}} = \frac{S}{\sqrt{n}}$$

Chi-Square

$$x^2 = \sum \frac{(o-e)^2}{e}$$

$\bar{x}$ = sample mean

n = size of the sample

s = sample standard deviation (i.e., the sample-based estimate of the standard deviation of the population)

o = observed results

e = expected results

Degrees of freedom are equal to the number of distinct possible outcomes minus one.

Chi-Square Table

p-value	Degrees of Freedom							
	1	2	3	4	5	6	7	8
0.05	3.84	5.99	7.82	9.49	11.07	12.59	14.07	15.51
0.01	6.64	9.21	11.34	13.28	15.09	16.81	18.48	20.09

Laws of Probability
If A and B are mutually exclusive, then:
$$P(A \text{ or } B) = P(A) + P(B)$$
If A and B are independent, then:
$$P(A \text{ and } B) = P(A) \times P(B)$$

Hardy-Weinberg Equations

$p^2 + 2pq + q^2 = 1$ p = frequency of the dominant allele in a population

$p + q = 1$ q = frequency of the recessive allele in a population

Metric Prefixes

Factor	Prefix	Symbol
10^9	giga	G
10^6	mega	M
10^3	kilo	k
10^{-2}	centi	c
10^{-3}	milli	m
10^{-6}	micro	μ
10^{-9}	nano	n
10^{-12}	pico	p

Mode = value that occurs most frequently in a data set

Median = middle value that separates the greater and lesser halves of a data set

Mean = sum of all data points divided by the number of data points

Range = value obtained by subtracting the smallest observation (sample minimum) from the greatest (sample maximum)

* For the purposes of the AP Exam, students will not be required to perform calculations using this equation; however, they must understand the underlying concepts and applications.

Rate and Growth

Rate

$$\frac{dY}{dt}$$

Population Growth

$$\frac{dN}{dt} = B - D$$

Exponential Growth

$$\frac{dN}{dt} = r_{max}N$$

Logistic Growth

$$\frac{dN}{dt} = r_{max}N\left(\frac{K-N}{K}\right)$$

Temperature Coefficient Q_{10} †

$$Q_{10} = \left(\frac{k_2}{k_1}\right)^{\frac{10}{T_2 - T_1}}$$

Primary Productivity Calculation

$$\frac{mg\ O_2}{L} \times \frac{0.698\ mL}{mg} = \frac{mL\ O_2}{L}$$

$$\frac{mL\ O_2}{L} \times \frac{0.536\ mg\ C\ fixed}{mL\ O_2} = \frac{mg\ C\ fixed}{L}$$

(at standard temperature and pressure)

dY = amount of change

dt = change in time

B = birth rate

D = death rate

N = population size

K = carrying capacity

r_{max} = maximum per capita growth rate of population

T_2 = higher temperature

T_1 = lower temperature

k_2 = reaction rate at T_2

k_1 = reaction rate at T_1

Q_{10} = the factor by which the reaction rate increases when the temperature is raised by ten degrees

Water Potential (ψ)

$\psi = \psi_P + \psi_S$

ψ_P = pressure potential

ψ_S = solute potential

The water potential will be equal to the solute potential of a solution in an open container because the pressure potential of the solution in an open container is zero.

The Solute Potential of a Solution

$\psi_S = -iCRT$

i = ionization constant (This is 1.0 for sucrose because sucrose does not ionize in water.)

C = molar concentration

R = pressure constant (R = 0.0831 liter bars/mole K)

T = temperature in Kelvin (C + 273)

Surface Area and Volume

Volume of a Sphere

$$V = \frac{4}{3}\pi r^3$$

Volume of a Rectangular Solid

$$V = lwh$$

Volume of a Right Cylinder

$$V = \pi r^2 h$$

Surface Area of a Sphere

$$A = 4\pi r^2$$

Surface Area of a Cube

$$A = 6a^2$$

Surface Area of a Rectangular Solid

$A = \Sigma$ surface area of each side

r = radius

l = length

h = height

w = width

s = length of one side of a cube

A = surface area

V = volume

Σ = sum of all

Dilution (used to create a dilute solution from a concentrated stock solution)

$C_i V_i = C_f V_f$

i = initial (starting) $\quad$ C = concentration of solute

f = final (desired) $\quad$ V = volume of solution

Gibbs Free Energy

$\Delta G = \Delta H - T\Delta S$

ΔG = change in Gibbs free energy

ΔS = change in entropy

ΔH = change in enthalpy

T = absolute temperature (in Kelvin)

$pH^* = -\log_{10}[H^+]$

* For the purposes of the AP Exam, students will not be required to perform calculations using this equation; however, they must understand the underlying concepts and applications.

† For use with labs only (optional)

NOTES

International Offices Listing

China (Beijing)
1501 Building A,
Disanji Creative Zone,
No.66 West Section of North 4th Ring Road Beijing
Tel: +86-10-62684481/2/3
Email: tprkor01@chol.com
Website: www.tprbeijing.com

China (Shanghai)
1010 Kaixuan Road
Building B, 5/F
Changning District, Shanghai, China 200052
Sara Beattie, Owner: Email: sbeattie@sarabeattie.com
Tel: +86-21-5108-2798
Fax: +86-21-6386-1039
Website: www.princetonreviewshanghai.com

Hong Kong
5th Floor, Yardley Commercial Building
1-6 Connaught Road West, Sheung Wan, Hong Kong
(MTR Exit C)
Sara Beattie, Owner: Email: sbeattie@sarabeattie.com
Tel: +852-2507-9380
Fax: +852-2827-4630
Website: www.princetonreviewhk.com

India (Mumbai)
Score Plus Academy
Office No.15, Fifth Floor
Manek Mahal 90
Veer Nariman Road
Next to Hotel Ambassador
Churchgate, Mumbai 400020
Maharashtra, India
Ritu Kalwani: Email: director@score-plus.com
Tel: + 91 22 22846801 / 39 / 41
Website: www.score-plus.com

India (New Delhi)
South Extension
K-16, Upper Ground Floor
South Extension Part–1,
New Delhi-110049
Aradhana Mahna: aradhana@manyagroup.com
Monisha Banerjee: monisha@manyagroup.com
Ruchi Tomar: ruchi.tomar@manyagroup.com
Rishi Josan: Rishi.josan@manyagroup.com
Vishal Goswamy: vishal.goswamy@manyagroup.com
Tel: +91-11-64501603/ 4, +91-11-65028379
Website: www.manyagroup.com

Lebanon
463 Bliss Street
AlFarra Building - 2nd floor
Ras Beirut
Beirut, Lebanon
Hassan Coudsi: Email: hassan.coudsi@review.com
Tel: +961-1-367-688
Website: www.princetonreviewlebanon.com

Korea
945-25 Young Shin Building
25 Daechi-Dong, Kangnam-gu
Seoul, Korea 135-280
Yong-Hoon Lee: Email: TPRKor01@chollian.net
In-Woo Kim: Email: iwkim@tpr.co.kr
Tel: + 82-2-554-7762
Fax: +82-2-453-9466
Website: www.tpr.co.kr

Kuwait
ScorePlus Learning Center
Salmiyah Block 3, Street 2 Building 14
Post Box: 559, Zip 1306, Safat, Kuwait
Email: infokuwait@score-plus.com
Tel: +965-25-75-48-02 / 8
Fax: +965-25-75-46-02
Website: www.scorepluseducation.com

Malaysia
Sara Beattie MDC Sdn Bhd
Suites 18E & 18F
18th Floor
Gurney Tower, Persiaran Gurney
Penang, Malaysia
Email: tprkl.my@sarabeattie.com
Sara Beattie, Owner: Email: sbeattie@sarabeattie.com
Tel: +604-2104 333
Fax: +604-2104 330
Website: www.princetonreviewKL.com

Mexico
TPR México
Guanajuato No. 242 Piso 1 Interior 1
Col. Roma Norte
México D.F., C.P.06700
registro@princetonreviewmexico.com
Tel: +52-55-5255-4495
+52-55-5255-4440
+52-55-5255-4442
Website: www.princetonreviewmexico.com

Qatar
Score Plus
Office No: 1A, Al Kuwari (Damas)
Building near Merweb Hotel, Al Saad
Post Box: 2408, Doha, Qatar
Email: infoqatar@score-plus.com
Tel: +974 44 36 8580, +974 526 5032
Fax: +974 44 13 1995
Website: www.scorepluseducation.com

Taiwan
The Princeton Review Taiwan
2F, 169 Zhong Xiao East Road, Section 4
Taipei, Taiwan 10690
Lisa Bartle (Owner): lbartle@princetonreview.com.tw
Tel: +886-2-2751-1293
Fax: +886-2-2776-3201
Website: www.PrincetonReview.com.tw

Thailand
The Princeton Review Thailand
Sathorn Nakorn Tower, 28th floor
100 North Sathorn Road
Bangkok, Thailand 10500
Thavida Bijayendrayodhin (Chairman)
Email: thavida@princetonreviewthailand.com
Mitsara Bijayendrayodhin (Managing Director)
Email: mitsara@princetonreviewthailand.com
Tel: +662-636-6770
Fax: +662-636-6776
Website: www.princetonreviewthailand.com

Turkey
Yeni Sülün Sokak No. 28
Levent, Istanbul, 34330, Turkey
Nuri Ozgur: nuri@tprturkey.com
Rona Ozgur: rona@tprturkey.com
Iren Ozgur: iren@tprturkey.com
Tel: +90-212-324-4747
Fax: +90-212-324-3347
Website: www.tprturkey.com

UAE
Emirates Score Plus
Office No: 506, Fifth Floor
Sultan Business Center
Near Lamcy Plaza, 21 Oud Metha Road
Post Box: 44098, Dubai
United Arab Emirates
Hukumat Kalwani: skoreplus@gmail.com
Ritu Kalwani: director@score-plus.com
Email: info@score-plus.com
Tel: +971-4-334-0004
Fax: +971-4-334-0222
Website: www.princetonreviewuae.com

Our International Partners

The Princeton Review also runs courses with a variety of partners in Africa, Asia, Europe, and South America.

Georgia
LEAF American-Georgian Education Center
www.leaf.ge

Mongolia
English Academy of Mongolia
www.nyescm.org

Nigeria
The Know Place
www.knowplace.com.ng

Panama
Academia Interamericana de Panama
http://aip.edu.pa/

Switzerland
Institut Le Rosey
http://www.rosey.ch/

All other inquiries, please email us at
internationalsupport@review.com